AS/A-LEVEL YEAR 1

STUDENT GUIDE

AQA

Chemistry

Physical chemistry 1

Alyn McFarland

Nora Henry

PHILIP ALLAN FOR
HODDER
EDUCATION
AN HACHETTE UK COMPANY

Philip Allan, an imprint of Hodder Education, an Hachette UK company, Blenheim Court, George Street, Banbury, Oxfordshire OX16 5BH

Orders

Bookpoint Ltd, 130 Milton Park, Abingdon, Oxfordshire OX14 4SB

tel: 01235 827827

fax: 01235 400401

e-mail: education@bookpoint.co.uk

Lines are open 9.00 a.m.–5.00 p.m., Monday to Saturday, with a 24-hour message answering service. You can also order through the Hodder Education website: www.hoddereducation.co.uk

© Alyn McFarland and Nora Henry 2015

ISBN 978-1-4718-4363-1

First printed 2015

Impression number 5 4 3 2 1

Year 2019 2018 2017 2016 2015

This guide has been written specifically to support students preparing for the AQA AS and A-level Chemistry examinations. The content has been neither approved nor endorsed by AQA and remains the sole responsibility of the author.

Cover photo: Ingo Bartussek/Fotolia

Typeset by Integra Software Services Pvt. Ltd, Pondicherry, India

Printed in Italy

Hachette UK's policy is to use papers that are natural, renewable and recyclable products and made from wood grown in sustainable forests. The logging and manufacturing processes are expected to conform to the environmental regulations of the country of origin.

Contents

Content Guidance

Questions & Answers

■ Getting the most from this book

Exam-style questions

Commentary on the questions

Tips on what you need to do to gain full marks, indicated by the icon (e)

Sample student answers

Practise the questions, then look at the student answers that follow.

Commentary on sample student answers

Find out how many marks each answer would be awarded in the exam and then read the comments (preceded by the icon (e)), which show exactly how and where marks are gained or lost.

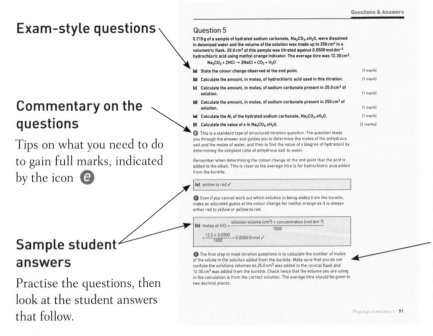

■About this book

This guide is the first of a series covering the AQA specifications for AS and A-level chemistry. It offers advice for the effective study of the physical chemistry sections 3.1.1 to 3.1.7, which are examined on AS papers 1 and 2 and also as part of A-level papers 1, 2 and 3 as shown in the table below.

Section	Topic	AS Paper 1	AS Paper 2	A-level Paper 1	A-level Paper 2	A-level Paper 3
3.1.1	Atomic structure	✓		✓		✓
3.1.2	Amount of substance	✓	✓	✓	✓	✓
3.1.3	Bonding	✓	✓	✓	✓	✓
3.1.4	Energetics	✓	✓	✓	✓	✓
3.1.5	Kinetics		✓		✓	✓
3.1.6	Chemical equilbria, Le Chatelier's principle and K_c	✓	✓	✓	✓	✓
3.1.7	Oxidation, reduction and redox equations	✓		✓		✓

The book has two sections:

■ The **Content Guidance** covers all the physical chemistry sections for AS and some of the A-level physical chemistry sections. It includes helpful tips on how to approach revision and improve exam technique. Do not skip over these tips as they provide important guidance. There are also knowledge check questions throughout this section, with answers at the end of the book. At the end of each section there is a summary of the key points covered.

■ The **Questions & Answers** section gives sample examination questions on each topic, as well as worked answers and comments on the common pitfalls to avoid.

The Content Guidance and Questions & Answers sections are divided into the topics listed on the AQA AS and A-level specifications.

Content Guidance

■ Atomic structure

Fundamental particles

Knowledge and understanding of atomic structure has evolved over time. Today we believe that an atom consists of a nucleus containing protons and neutrons surrounded by electrons moving in shells (energy levels). The properties of the three fundamental particles of an atom are shown in Table 1.

Subatomic particle	Relative mass	Relative charge	Location in atom
Proton	1	+1	Nucleus
Neutron	1	0	Nucleus
Electron	1/1840	−1	Shells

Table 1

Note that the masses and charges of these particles are so small that it is easier to use standard measures and compare the rest to them — hence the term 'relative' is used.

Mass number and isotopes

Calculating the number of fundamental particles in atoms and ions

- The **atomic number** (proton number) (Z) is equal to the number of protons in the nucleus of an atom.
- The **mass number** (A) is the total number of protons and neutrons in the nucleus of an atom.

These numbers can be used to calculate the number of fundamental particles in an atom of an element.

number of protons = atomic number

number of neutrons = mass number − atomic number

Table 2 shows some examples.

Element	Mass number (A)	Atomic number (Z)	Number of protons	Number of electrons	Number of neutrons
H	1	1	1	1	1 − 1 = 0
Al	27	13	13	13	27 − 13 = 14
Br	79	35	35	35	79 − 35 = 44

Table 2

Exam tip

The symbol for an element may be written with its atomic number and mass number, i.e. $^A_Z X$, where X is the symbol for the element, Z is the atomic number and A is the mass number.

Atoms are electrically neutral as they have equal numbers of protons and electrons. Simple ions are charged particles formed when atoms lose or gain electrons. When an atom *loses electrons* it becomes a **positive ion**, when an atom *gains electrons* it becomes a **negative ion.** The number of electrons subtracted from the number of protons gives the charge. Remember that an atom has no overall charge.

charge = number of protons − number of electrons

Worked example

Determine the number of fundamental particles present in an aluminium ion, Al^{3+}.

Answer

From the periodic table, the atomic number of aluminium is 13 and the relative atomic mass is 27.0. Apart from chlorine (relative atomic mass 35.5), the relative atomic mass can be taken as the mass number of the most common isotope.

number of protons = atomic number = 13

number of neutrons = mass number − atomic number = 27 − 13 = 14

charge = number of protons − number of electrons

+3 = 13 − number of electrons

number of electrons = 13 − (+3) = 10

Isotopes

Isotopes of an element are atoms with the same number of protons but a different number of neutrons. Hence isotopes have the same atomic number but a different mass number.

Isotopes of an element have the same chemical properties — this is because they have the same number of electrons, and the same number of electrons in the outer shell.

Time of flight (TOF) mass spectrometer

A mass spectrometer is an analytical instrument used to give accurate information about relative isotopic mass and relative abundance of isotopes. It can also be used to help identify elements and to determine relative molecular mass. The five processes that occur in a TOF mass spectrometer are:

1 **Electrospray ionisation**. A high voltage is applied to the sample (which is dissolved in a polar volatile solvent). Molecules (or atoms) lose an electron and form gaseous positive ions. Molecules must be ionised so they can be accelerated and detected.

2 **Acceleration**. The cations are accelerated to a constant kinetic energy by passing through an electric field; they move towards a negative plate. The speed the ions reach depends on their mass — the lighter the ions the faster they travel.

3 **Ion drift**. This occurs in the flight tube where there is no electric field and the ions are separated based on their speed; the smaller, fast ions travel through the flight tube most rapidly and reach the detector first.

Exam tip

It is the atomic number that defines the identity of the particle. A particle with 17 protons is always a chlorine particle — it may be a chlorine atom or a chloride ion depending on the number of electrons.

Knowledge check 1

Give the symbol of the element that has an isotope with a mass number of 56 and has 30 neutrons in its nucleus.

4 **Ion detection**. When the cations reach the detector they cause a small current due to their charge. The detector records the different flight times of the ions.

5 **Data analysis**. The flight times are analysed and recorded as a mass spectrum. A mass spectrum is a plot of relative abundance against mass to charge ratio (*m/z*).

Interpretation of the mass spectrum of an element

The mass spectrum of an element gives accurate information about relative isotopic mass and also about relative abundance of isotopes. This information can be used to calculate the relative atomic mass of the element.

Relative isotopic masses

The relative isotopic mass is the mass of a single isotope of an element relative to the mass of an atom of carbon-12.

The mass spectrometer detects individual ions, so different isotopes are detected because they have different relative isotopic masses. The mass spectrum shown in Figure 1 has four peaks and hence it has four isotopes with relative isotopic masses of 50, 52, 53 and 54.

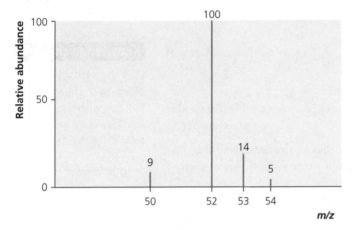

Figure 1 TOF mass spectrum of an element

Abundances

The abundance of each isotope is given by the height of the peak. In Figure 1 the numbers above the peaks indicate the relative abundance of each ion detected. The ion $^{52}M^+$, with *m/z* 100, has the highest abundance.

Calculation of relative atomic mass and identification of elements

The relative atomic mass (A_r) of an element by definition is given by:

$$A_r = \frac{\text{average mass of 1 atom of an element}}{\frac{1}{12} \text{ mass of one atom of } ^{12}C}$$

The relative atomic mass of an element can be calculated from the relative isotopic masses of the isotopes (which are the same as the mass numbers) and the relative proportions in which they occur.

> **Exam tip**
>
> If you are asked to outline how the TOF mass spectrometer is able to separate two ions with different *m/z* values to give two peaks you must mention how the ions are accelerated by the electric field to constant kinetic energy and that the ion with a smaller mass moves faster and arrives at the detector first.

$$A_r = \frac{\sum (\text{mass of isotopes} \times \text{relative abundance})}{\sum (\text{relative abundance})}$$

where $\sum$ represents the 'sum of' for all isotopes.

For the spectrum shown in Figure 1:

$$A_r = \frac{(50 \times 9) + (52 \times 100) + (53 \times 14) + (54 \times 5)}{9 + 100 + 14 + 5} = \frac{6662}{128} = \frac{52.0 \text{ (to 1}}{\text{decimal place)}}$$

You may be asked to identify the element that produced this mass spectrum. Remember to look at the relative atomic mass (A_r) and *not* the atomic number. Chromium has a relative atomic mass of 52.0 — writing tellurium (atomic number 52) in this case is a common error.

You may be also asked to identify the ion responsible for a peak — for example, what ion causes the peak at 52? The peak at 52 is caused by the ion $^{52}Cr^+$.

Worked example 1

A sample of boron contains the isotopes ^{10}B and ^{11}B only. The isotope ^{11}B has abundance four times greater than that of ^{10}B. Calculate the relative atomic mass of boron in this sample. Give your answer to 1 decimal place.

Answer

In the absence of actual values of abundance use the ratio. ^{11}B has an abundance four times greater than that of ^{10}B so let the abundance of ^{11}B equal 4 and the abundance of ^{10}B equal 1.

$$A_r = \frac{(11 \times 4) + (10 \times 1)}{4 + 1} = \frac{54}{5} = 10.8$$

Worked example 2

A mass spectrum of a sample of indium showed two peaks at $m/z = 113$ and $m/z = 115$. The relative atomic mass of this sample of indium is 114.5. Use these data to calculate the ratio of the relative abundances of the two isotopes.

Answer

In this case you are asked for the abundances. Let the abundance of ^{113}In be x and the abundance of ^{115}In be y. Carry out the calculation in the same way, multiplying the mass by the abundance and dividing by the total abundance.

$$A_r = 114.5 = \frac{113x + 115y}{x + y}$$

$$114.5(x + y) = 113x + 115y$$

$$114.5x + 114.5y = 113x + 115y$$

$$114.5x - 113x = 115y - 114.5y$$

$$1.5x = 0.5y$$

Ratio of abundances is $3 : 1$.

Exam tip

A common mistake is to use the incorrect denominator. Make sure that you find it by totalling the abundances; do not assume that this value is 100 — it is only 100 if percentage abundances are used. Often you are asked to give your answer to a certain number of decimal places. A_r values are usually asked for to 1 decimal place.

Mass spectrum of molecules

For a compound the mass spectrum is more complicated as the molecule breaks up and produces smaller ions called fragments, which also give peaks. The relative molecular mass of the compound can be determined by looking at the peak with the largest m/z value — the **molecular ion** peak. The molecular ion is the ion formed by the removal of one electron from a molecule. The mass value for the molecular ion peak is the same as the relative molecular mass of the compound.

Figure 2 shows the mass spectrum of a molecule of ethanol. The species responsible for the peak at 46 is $CH_3CH_2OH^+$.

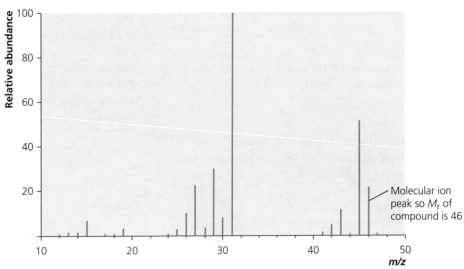

Figure 2 TOF mass spectrum for ethanol, CH_3CH_2OH

Electron configuration

Electrons are arranged in energy levels (the equivalent of shells) in which the energy of the electrons increases with increasing distance from the nucleus. The energy levels are labelled $n = 1$ (closest to the nucleus), $n = 2$, $n = 3$ etc.

Energy levels (shells) are subdivided into subshells, designated s, p, d (and f), which are made up of orbitals. Two electrons can occupy each orbital. The different subshells can each hold a different maximum number of electrons, as shown in Table 3.

Subshell	Maximum number of electrons	Number of orbitals in the subshell
s	2	1
p	6	3
d	10	5
f	14	7

Table 3

The first shell ($n = 1$) contains only an s subshell, at $n = 2$ there is an s subshell and a p subshell and at $n = 3$ there is an s subshell, a p subshell and a d subshell.

Exam tip

If you are given a mass spectrum for a compound and asked for the relative molecular mass, simply give the m/z value of the peak furthest to the right — the molecular ion peak.

Exam tip

You must be able to write electron configurations of atoms and ions with mass number up to 36. You are not required to write electron configurations using the f subshell.

Electrons fill up the lowest energy subshells first. Note that **the 4s subshell is lower in energy than the 3d** so it fills before $3d$. Hence the subshells fill in the following order:

1s 2s 2p 3s 3p 4s 3d 4p

but the subshells should be written in the following order:

1s 2s 2p 3s 3p 3d 4s 4p

When atoms form positive ions, electrons are lost from subshells in the following order:

4p 4s 3d 3p 3s 2p 2s 1s

The 2+ ion for transition metals is the most common and this is caused by loss of $4s^2$ electrons (not $3d$).

When atoms form negative ions, electrons are added to the highest energy occupied subshell.

Determining electron configuration of atoms and ions

An iron atom
- Atomic number of iron = 26, so an iron atom has 26 protons.
- Atoms are electrically neutral, so an atom of iron has 26 electrons.
- Using order of filling: $1s^2\, 2s^2\, 2p^6\, 3s^2\, 3p^6\, 4s^2\, 3d^6$.
- Remember that when writing the configuration, the $4s$ has to come after the $3d$.
- So the electron configuration of an iron atom is: $1s^2\, 2s^2\, 2p^6\, 3s^2\, 3p^6\, 3d^6\, 4s^2$.

An iron(II) ion
- The electron configuration of iron atom: $1s^2\, 2s^2\, 2p^6\, 3s^2\, 3p^6\, 3d^6\, 4s^2$.
- Remember that transition metal atoms lose their $4s$ electrons first.
- An iron atom has lost two electrons to form a 2+ ion.
- So the electron configuration of an iron(II) ion is: $1s^2\, 2s^2\, 2p^6\, 3s^2\, 3p^6\, 3d^6$.

A bromide ion, Br⁻
- The electron configuration of a bromine atom is determined first.
- The atomic number of bromine = 35 so a Br atom has 35 protons and 35 electrons.
- The electron configuration of Br atom: $1s^2\, 2s^2\, 2p^6\, 3s^2\, 3p^6\, 3d^{10}\, 4s^2\, 4p^5$.
- A bromine atom has gained one electron to form a bromide ion.
- The electron configuration of a bromide ion is: $1s^2\, 2s^2\, 2p^6\, 3s^2\, 3p^6\, 3d^{10}\, 4s^2\, 4p^6$.

Chromium and copper atoms
The electron configurations of chromium and copper atoms are unusual:

Cr: $1s^2\, 2s^2\, 2p^6\, 3s^2\, 3p^6\, 3d^5\, 4s^1$ (*not* $3d^4\, 4s^2$)

Cu: $1s^2\, 2s^2\, 2p^6\, 3s^2\, 3p^6\, 3d^{10}\, 4s^1$ (*not* $3d^9\, 4s^2$)

Cr and Cu have unusual electron configurations owing to the stability of the half filled and filled d^5 and d^{10} configurations. However, the formation of ions of copper and chromium works in the normal way with the loss of the $4s$ electrons first.

Exam tip

Remember that transition metal atoms lose their 4s electrons first. A common question is to ask for the electron configuration of a transition metal ion.

Knowledge check 2

Identify the element that has a 2+ ion with an electron configuration of $1s^2\, 2s^2\, 2p^6\, 3s^2\, 3p^6\, 3d^{10}$.

Exam tip

As soon as you start any exam, put a star (*) at Cr and Cu on your periodic table to remind you that their electron configurations are different from what you would expect.

'Electron-in-box' diagrams

To help understanding, it is useful to be able to write an electron configuration in an 'electron-in-box' form (Figure 3). When filling the boxes remember that:

- Electrons occupy the lowest available energy level.
- Electrons only pair when no other space is available in the subshell — hence electrons do not pair up until a subshell is half filled.
- In each single paired orbital, electrons spin in opposite directions to minimise repulsions. Opposite spin is represented as: ↑↓ in 'electron-in-box' diagrams.
- Electrons in a subshell that are not paired spin in the same direction and are represented by arrows pointing in the same direction.
- Note that the $4s$ subshell is slightly lower in energy than the $3d$ and is filled first. However, when forming positive ions, $4s$ electrons are lost first.

Figure 3 Electron-in-box notation

Figure 3 shows three p subshells, each containing three orbitals, with different arrangements of electrons. A d subshell is also shown, split into five orbitals, each one able to hold two electrons.

Worked examples

A nitrogen atom (atomic number 7)

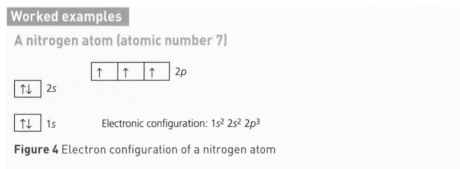

Figure 4 Electron configuration of a nitrogen atom

An iron(III) ion (atomic number 26)

The electron configuration of an iron atom is: $1s^2\ 2s^2\ 2p^6\ 3s^2\ 3p^6\ 3d^6\ 4s^2$

The electron configuration of an iron(III) ion is: $1s^2\ 2s^2\ 2p^6\ 3s^2\ 3p^6\ 3d^5$

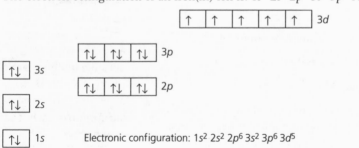

Figure 5 Electron configuration of an iron(III) ion

Exam tip

Remember that electrons do not pair up until a subshell is half filled. Note that unpaired electrons spin in the same direction, as shown by the arrows pointing in the same direction.

Ionisation energy

- **First ionisation energy** is the energy required to remove 1 mole of electrons from 1 mole of gaseous atoms to form 1 mole of gaseous 1+ ions.
- **Second ionisation energy** is the energy required to remove 1 mole of electrons from 1 mole of gaseous 1+ ions to form 1 mole of gaseous 2+ ions.
- **Third ionisation energy** is the energy required to remove 1 mole of electrons from 1 mole of gaseous 2+ ions to form 1 mole of gaseous 3+ ions.

Values of ionisation energies are always endothermic, as heat is needed to overcome the attraction between the negative electron and the positive nucleus. It is measured in $kJ\,mol^{-1}$.

In the following examples, X represents any element.

Equation for first ionisation energy:

$$X(g) \rightarrow X^+(g) + e^-$$

Equation for second ionisation energy:

$$X^+(g) \rightarrow X^{2+}(g) + e^-$$

Equation for third ionisation energy:

$$X^{2+}(g) \rightarrow X^{3+}(g) + e^-$$

Trend in ionisation energy down a group

Down a group the first ionisation energy *decreases* because:

- there is an increase in atomic radius
- there is more shielding of the outer electron from the nuclear charge due to increased number of shells
- there is less nuclear attraction for the outer electron because of the increased atomic radius and increased shielding

Figure 6 shows the decrease in first ionisation energy down group 2.

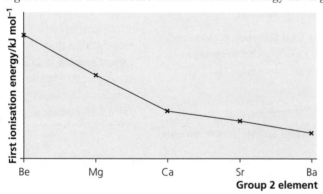

Figure 6 First ionisation energies of group 2 elements

Trend in ionisation energy across a period

Across a period the first ionisation energy shows a general *increase* because:

- there is an increase in nuclear charge across the period
- the shielding is similar as the electron is being removed from the same shell

■ there is a smaller atomic radius, as the outermost electron is held closer to the nucleus by the greater nuclear charge

Figure 7 shows the general increase in first ionisation energy across a period. Notice that there are some elements that deviate from the general trend. For Al (group 3) and S (group 6) the first ionisation energy value drops below the general increase.

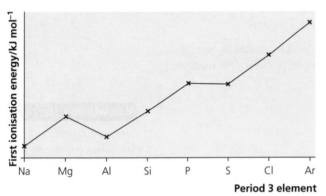

Figure 7 First ionisation energies of period 3 elements

Group 3

Magnesium atom: $1s^2\, 2s^2\, 2p^6\, 3s^2$

Aluminium atom: $1s^2\, 2s^2\, 2p^6\, 3s^2\, 3p^1$

The first ionisation energy for aluminium is lower than expected because the $3p^1$ electron is further from the nucleus and has more shielding from the inner $3s^2$ electrons, so it requires less energy to remove it.

This pattern in ionisation energy provides evidence for the existence of subshells.

Group 6

Phosphorus atom: $1s^2\, 2s^2\, 2p^6\, 3s^2\, 3p^3$ has a half filled $3p$ subshell and is stable.

Sulfur atom: $1s^2\, 2s^2\, 2p^6\, 3s^2\, 3p^4$ has a more than half filled $3p$ subshell.

The lower than expected first ionisation energy for sulfur (and all group 6 elements) is due to the pairing of electrons in the p subshell — the repulsion between them lowers the energy required to remove one of the electrons and decreases the first ionisation energy.

Successive ionisation energies

Successive ionisation energies get larger. This is because when an electron is removed in a first ionisation, a positive ion is left.

$$Mg(g) \rightarrow Mg^+(g) + e^-$$

To remove an electron from a positive ion requires more energy as the outermost electron is closer to the nucleus and there is the same number of protons for fewer electrons, resulting in a greater nuclear attraction for the electrons.

Often, a question may give successive ionisation energies for some elements and ask you to determine the group of the periodic table to which the elements belong. Look for the large jump in successive ionisation energies. This jump occurs after all the outer energy level electrons have been removed and will indicate the number of electrons in the outer energy level.

Worked example

An element has successive ionisation energies as shown in Table 4. To which group of the periodic table does it belong?

Ionisation energy/ kJ mol^{-1}	First	Second	Third	Fourth	Fifth	Sixth
	580	1800	2700	11 600	14 800	18 400

Table 4

Answer

The large jump in successive ionisation energy values occurs after three electrons are removed so there must be three electrons in the outer energy level, which would indicate a group 3 element.

Summary

- Atoms consist of three fundamental particles — protons (+ charge) and neutrons (0 charge) located in the nucleus, and electrons (– charge) found in shells. Electrons have a mass of 1/1840 that of a proton or neutron.
- To calculate the number of fundamental particles in an atom use the equations:

  number of protons = atomic number

  number of neutrons = mass number – atomic number

  charge on an ion = number of protons – number of electrons

- TOF mass spectrometry involves electrospray ionisation, acceleration to give all ions constant kinetic energy, ion drift, ion detection and data analysis. For an element, mass spectrometry measures the mass and relative abundance of each isotope.
- Relative atomic mass can be calculated by multiplying each relative isotopic mass by each abundance and dividing the total by the total abundance. The relative molecular mass is given by the highest m/z value on the spectrum of a molecule.
- The first ionisation energy is the energy required to remove 1 mole of electrons from 1 mole of gaseous atoms to form 1 mole of gaseous 1+ ions.
- Ionisation energies generally increase across a period due to increased nuclear charge, similar shielding and decreased atomic radius, but decrease down a group due to increased atomic radius and shielding.

Amount of substance

Writing chemical formulae

Writing chemical formulae is an important skill and mistakes in equations are commonly made due to incorrect formulae. Many covalent compounds that you come across are simply learned or can be worked out from the name or, in organic chemistry, the formulae are determined from a general formula.

Formulae of ionic compounds from the ions

The formula of an ionic compound can be determined simply from the charges on the ions as the overall charge on an ionic compound must be zero. For example:

1 Sodium chloride contains Na^+ and Cl^- ions. One of each ion is required, so sodium chloride is NaCl.

2 Calcium chloride contains Ca^{2+} and Cl^- ions. Two Cl^- ions are required for one Ca^{2+} ion, so calcium chloride is $CaCl_2$.

3 Magnesium oxide contains Mg^{2+} and O^{2-} ions. One of each ion is required, so magnesium oxide is MgO.

Exam tip

The charge on simple ions such as sodium ions and chloride ions can be worked out from the group number. Metals form positive ions and non-metals form negative ions.

- group 1 elements form 1+ ions; for example, Na^+
- group 2 elements form 2+ ions; for example, Mg^{2+}
- group 3 metallic elements form 3+ ions; for example, Al^{3+}
- group 5 non-metallic elements form 3– ions; for example, N^{3-}
- group 6 non-metallic elements form 2– ions, for example, O^{2-}
- group 7 elements form 1– ions, for example, Cl^-

Compound ions are charged particles made up of more than one atom. Common examples are shown in Table 5.

Ion	Formula	Ion	Formula
sulfate	$SO_4{}^{2-}$	carbonate	$CO_3{}^{2-}$
nitrate	$NO_3{}^-$	hydrogen carbonate	$HCO_3{}^-$
nitrite	$NO_2{}^-$	sulfite	$SO_3{}^{2-}$
hydrogen sulfate	$HSO_4{}^-$	hypochlorite	OCl^-
dichromate(vi)	$Cr_2O_7{}^{2-}$	manganate(vii)	$MnO_4{}^-$
ammonium	$NH_4{}^+$	hydroxide	OH^-

Table 5

Remember that chemical formulae are written without the charges.

4 Copper(ii) hydroxide contains Cu^{2+} and OH^- ions. Two OH^- ions are required for one Cu^{2+} ion, so copper(ii) hydroxide is $Cu(OH)_2$.

Exam tip

Mistakes are commonly encountered in the formulae for the hydroxides of metal ions with charge greater than 1+, in the formulae for the carbonates and sulfates of metal ions with charge of 1+, and in those for ammonium carbonate and ammonium sulfate.

- Commonly, brackets are left out, for example, calcium hydroxide is often **incorrectly** written as $CaOH_2$, instead of correctly as $Ca(OH)_2$.
- The two metal or ammonium ions may not be included in the formula, for example, potassium sulfate is often **incorrectly** written as KSO_4 instead of correctly as K_2SO_4.

Examples of correct sulfate and carbonate formulae are Na_2CO_3, Na_2SO_4, K_2CO_3, $(NH_4)_2SO_4$.

Exam tip

You need to learn the formulae (including charges) of common compound ions.

Exam tip

The hydroxide ion is a unit and so when we have more than one of them we must use brackets. This is true of all compound ions.

Knowledge check 6

Write the chemical formula for ammonium dichromate(vi).

Relative atomic mass and relative molecular mass

Relative atomic mass (A_r) is the average mass of an atom of an element relative to $\frac{1}{12}$ the mass of an atom of carbon-12.

The relative molecular mass (M_r) is the total of the relative atomic masses for all atoms in a compound. RFM (relative formula mass) is often used for ionic compounds but M_r is usually used for all these values.

1 mole of a substance is the M_r measured in grams.

1 mole of Mg = 24.3 g	(A_r of Mg = 24.3)
1 mole of N_2 = 28.0 g	(A_r of N = 14.0)
1 mole of $Fe(NO_3)_3$ = 241.8 g	(A_r of N = 14.0; A_r of O = 16.0; A_r of Fe = 55.8)

The term molar mass is also used, which means the mass of 1 mole. The units of molar mass are g/mol or $g\,mol^{-1}$.

Molar mass of $Cu(OH)_2$ = 97.5 $g\,mol^{-1}$

Molar mass of N_2 = 28.0 $g\,mol^{-1}$

The mole and the Avogadro constant

A balanced symbol equation for a reaction gives the rearrangement of the atoms within a chemical reaction.

Using the equation

$$Cu + S \rightarrow CuS$$

we can read this as one Cu atom reacts with one S atom to form one CuS unit.

However, the masses of atoms are too small to be measured so the number of particles used in measurements is scaled up by 6.02×10^{23}. This number is called the **Avogadro constant** and is often represented by the symbol L.

For example, one Cu atom has a mass of 1.055×10^{-22} g, so 6.02×10^{23} Cu atoms have a mass of approximately 63.5 g.

The term 'amount' is the quantity that is measured in moles. Mole is written as mol for unit purposes. For example, the equation

$$Cu + S \rightarrow CuS$$

can be read as 1 mole of copper atoms reacts with 1 mole of sulfur atoms to form 1 mole of copper(II) sulfide.

Calculating amount, in moles

The amount of a substance is measured in moles and is often represented by n. It is calculated from the mass using the expression:

amount, in moles, $n = \dfrac{\text{mass (g)}}{M_r}$

The **Avogadro constant** is defined as the number of atoms in 12.000 g of carbon-12.

The amount of a substance that contains the Avogadro constant (6.02×10^{23}) particles (atoms, molecules or groups of ions) is called a **mole** of the substance.

The expression can be rearranged to calculate mass in grams from the amount in moles (n) and M_r:

mass (g) $= n \times M_r$

Or M_r may be calculated from mass and amount in moles (n):

$$M_r = \frac{\text{mass (g)}}{n}$$

Worked example

Calculate the amount, in moles, present in 2.71 g of carbon dioxide. Give your answer to 3 significant figures.

Answer

$$\text{amount, in moles} = \frac{\text{mass (g)}}{M_r} = \frac{2.71}{44.0} = 0.0616 \, \text{mol}$$

Calculating amount, in moles, in a solution

The concentration of any solution is measured in mol dm^{-3}. A solution of concentration $1 \, \text{mol dm}^{-3}$ will have 1 mol of solute dissolved in $1 \, \text{dm}^3$.

The amount, in moles, of a solute in a solution may be calculated using the expression:

$$\text{amount, in moles} \ (n) = \frac{\text{volume} \left(\text{cm}^3\right) \times \text{concentration} \left(\text{mol dm}^{-3}\right)}{1000}$$

Worked example

Calculate the amount, in moles, of sodium hydroxide present in 15.0 cm³ of a solution of concentration $1.25 \, \text{mol dm}^{-3}$. Give your answer to 3 significant figures.

Answer

$$\text{amount, in moles} \ (n) = \frac{\text{volume} \left(\text{cm}^3\right) \times \text{concentration} \left(\text{mol dm}^{-3}\right)}{1000}$$

$$= \frac{15.0 \times 1.25}{1000} = 0.0188 \, \text{mol}$$

Using the Avogadro constant

Calculations involving number of particles or mass of a certain number of particles will use the Avogadro constant (L), which is equal to 6.02×10^{23}.

$$\frac{\text{mass (g)}}{M_r} = \text{amount in moles} = \frac{\text{number of particles}}{L}$$

These terms can be rearranged to give:

mass (g) $= \text{moles} \times M_r$

and

number of particles $= \text{moles} \times L$

Exam tip

Units are very important in chemistry. $1 \, \text{dm}^3$ is the same as 1 litre but chemists prefer $1 \, \text{dm}^3$. $1 \, \text{cm}^3$ is the same as 1 millilitre. There are $1000 \, \text{cm}^3$ in $1 \, \text{dm}^3$.

Exam tip

Both the examples above asked for the answer to be given to 3 significant figures. This is to be expected when all the initial data were given to 3 significant figures. Always give a numerical answer to the number of significant figures required by the question.

Knowledge check 7

Calculate the amount, in moles, of potassium carbonate present in 2.08 g. Give your answer to 3 significant figures.

Calculate the number of hydrogen atoms present in 0.125 g of methane, CH_4.

Answer

$$\frac{mass(g)}{M_r} = \frac{0.125}{16.0} = 7.813 \times 10^{-3}\,mol \times L\,(6.02 \times 10^{23})$$

$$= 4.70 \times 10^{21}\text{ molecules of }CH_4$$

Each CH_4 contains four H atoms, so:

number of H atoms $= 4 \times 4.70 \times 10^{21}$

$$= 1.88 \times 10^{22}\text{ atoms of H}$$

Calculate the number of atoms in 1.00 g of argon gas. Give your answer to 3 significant figures.

The ideal gas equation

The ideal gas equation is:

$$pV = nRT$$

where p is pressure measured in pascals (Pa), V is volume measured in m^3, n is the amount of the substance in moles, R is the gas constant, which is given as $8.31\,J\,K^{-1}\,mol^{-1}$, and T is the temperature in kelvin (K).

Calculations using this expression often involve the conversion of pressure and volumes and sometimes temperature from different units.

In the reaction

$$4LiNO_3(s) \rightarrow 2Li_2O(s) + 4NO_2(g) + O_2(g)$$

2.17 g of sodium nitrate were heated until constant mass. Calculate the total volume of gas, in dm^3, produced at 298 K at 110 kPa. (The gas constant $R = 8.31\,J\,K^{-1}\,mol^{-1}$.)

Answer

amount, in moles, of $LiNO_3 = \dfrac{2.17}{68.9} = 0.0315\,mol$

4 moles of $LiNO_3$ produces 5 moles of gas: $4NO_2(g) + O_2(g)$

0.0315 mol of $LiNO_3$ produces 0.0394 mol of gas

$pV = nRT$

$p = 110\,000\,Pa$, $V = ?$, $n = 0.0394\,mol$, $R = 8.31\,J\,K^{-1}\,mol^{-1}$, $T = 298\,K$

$$V = \frac{nRT}{p} = \frac{0.0394 \times 8.31 \times 298}{110\,000} = 0.000887\,m^3$$

$V = 0.887\,dm^3$

The key to any type of question like this is making sure that the pressure, volume and temperature are in correct units to match the gas constant. The gas constant is in $J\,K^{-1}\,mol^{-1}$, so pressure is in Pa, volume in m^3 and temperature in K.

Temperature may be given in °C. Convert temperature in °C to K by adding 273. For example, 100°C is 373 K.

Molar gas volume

At 25°C ($T = 298\,K$), atmospheric pressure ($p = 101\,325\,Pa$) and using a gas constant, $R = 8.31\,J\,K^{-1}\,mol^{-1}$, the volume of 1 mole ($n = 1$) of a gas is calculated as:

$$pV = nRT$$

$$V = \frac{nRT}{p} = \frac{1 \times 8.31 \times 298}{101325} = 0.0244\,m^3$$

$$V = 24.4\,dm^3$$

Often $24\,dm^3$ is used as the molar gas volume for 1 mole of any gas at 25°C and atmospheric pressure.

Empirical and molecular formulae

Types of formulae

There are two main types of chemical formulae:

■ An **empirical formula** shows the simplest ratio of the atoms of each element. This type of formula is used for ionic compounds and macromolecules (giant covalent molecules).
Examples: NaCl (ionic); MgO (ionic); $CaCl_2$ (ionic); SiO_2 (macromolecular)

■ A **molecular formula** shows the actual number of atoms of each element in one molecule of the substance.
This type of formula is used for all molecular (simple) covalent substances.
Examples: H_2O; CO_2; O_2; CH_4; NH_3; H_2O_2; I_2; S_8 (all molecular covalent)

Some elements exist as simple molecules. These are the diatomic elements (H_2, N_2, O_2, F_2, Cl_2, Br_2, I_2), sulfur (S_8) and phosphorus (P_4).

The empirical formula states the simplest ratios of all the elements in the compound. For example, ethane has molecular formula C_2H_6 but its empirical formula is CH_3.

The M_r of a substance will be the same as the mass of the atoms in the empirical formula if the empirical formula is the same as the molecular formula, or it will be a simple multiple.

The term molar mass may be used in place of M_r.

The molar mass is the mass of 1 mole and is measured in units of $g\,mol^{-1}$.

Worked example

The empirical formula of a compound is CH. The M_r is determined to be 78.0. What is the molecular formula of the compound?

Answer

empirical formula: CH, empirical formula mass: 13.0, M_r: 78.0

$$\frac{\text{molar mass}}{\text{empirical formula mass}} = \frac{78.0}{13.0} = 6$$

So:

molecular formula = 6 × empirical formula

molecular formula = C_6H_6

Finding formulae

The empirical formula of a compound may be determined from mass measurements taken during a reaction between two elements or from heating hydrated compounds to constant mass to remove the water of crystallisation. The formula can also be determined from the given percentage composition by mass.

Worked example

0.91 g of titanium combined with oxygen to give 1.52 g of an oxide of titanium. Find the formula of the oxide of titanium.

Practically, this is done by heating a certain mass of titanium in a crucible with a lid, which is raised periodically to let fresh air in. The titanium is heated to constant mass to ensure that all the titanium has combined to form the oxide.

1 Find the mass of the empty crucible: 16.18 g
2 Find the mass of the crucible and some titanium: 17.09 g
3 Mass of titanium = (2) − (1) = 17.09 − 16.18 = 0.91 g
4 Find the mass of the crucible after heating to constant mass: 17.70 g
5 Mass of oxygen combined = (4) − (2) = 17.70 − 17.09 = 0.61 g

Answer

Using this information we can now calculate the formula of the oxide of titanium, as shown in Table 6.

Element	Titanium	Oxygen
Mass (g)	0.91	0.61
A_r	47.9	16.0
Moles	0.91/47.9 = 0.019	0.61/16.0 = 0.038
Ratio	1	2
Formula	TiO_2	

Table 6

As the ratio is one Ti atom to two O atoms, we say that the *empirical formula* is TiO_2, but it could also be Ti_2O_4 or Ti_3O_6, etc., as the ratio in these compounds is the same.

If the moles work out to be not as simple as the ones shown in the example, then to calculate the ratio, divide all the moles by the smallest number of moles. If you end up with, say, 0.5, then multiply all the ratio numbers by 2 to get whole numbers.

Hydrated salts

Many salts (formed from acids) are hydrated when they are solid. A hydrated salt contains **water of crystallisation**.

If hydrated salts are heated to constant mass in an open container (so the water vapour can escape) all of the water of crystallisation is removed, and an **anhydrous salt** remains.

Hydrated salts are written with the water of crystallisation, for example, $CuSO_4.5H_2O$, or $CoCl_2.6H_2O$, or $Na_2CO_3.10H_2O$.

Exam tip

Use the A_r of oxygen as 16.0 as we are dealing with oxygen atoms combined in the formula.

Knowledge check 11

What is the empirical formula of the chloride of mercury when 1.20 g of mercury combines with 0.425 g of chlorine? Determine the empirical formula of the chloride of mercury.

Water of crystallisation refers to water molecules chemically bonded within a crystal structure.

An **anhydrous salt** contains no water of crystallisation.

Content Guidance

The number of moles of water of crystallisation attached to 1 mole of the salt is called the degree of hydration. Many hydrated salts effloresce when left in the open. This means they lose their water of crystallisation gradually to the atmosphere. Heating in an open container removes the water of crystallisation more rapidly. Heating a hydrated salt to constant mass will remove all of the water of crystallisation.

The degree of hydration can be determined by taking mass measurements before heating and after heating to constant mass.

Worked example

3.48 g of a sample of hydrated sodium carbonate, $Na_2CO_3.nH_2O$, produces 1.59 g of the anhydrous sodium carbonate, Na_2CO_3, on heating to constant mass. Find the value of n in the formula of the hydrated salt.

Answer

mass of hydrated salt = 3.48 g

mass of anhydrous salt = 1.59 g

mass of water lost = 3.48 − 1.59 = 1.89 g

Table 7 shows how the value of n can be calculated

Compound	Sodium carbonate	Water
Formula	Na_2CO_3	H_2O
Mass (g)	1.59 g	1.89 g
M_r	106.0	18.0
Moles	1.59/106.0 = 0.015	1.89/18.0 = 0.105
Ratio (÷0.015)	1	7
Empirical formula	$Na_2CO_3.7H_2O$	

Table 7

We can see from the empirical formula that the value of $n = 7$.

In the type of calculation shown in the worked example, the M_r of the hydrated salt may be determined instead. The number of moles of the hydrated salt is the same as the number of moles of the anhydrous salt.

So, 0.015 mol of Na_2CO_3 formed from 0.015 mol of $Na_2CO_3.nH_2O$.

Using the mass of the hydrated salt (3.48 g) and the number of moles (0.015 mol), the M_r is calculated as 3.48/0.015 = 232.0.

Subtracting the mass of the Na_2CO_3 (106.0) leaves 126.0.

This 126.0 is due to the mass of the water, so divide by the M_r of water (18.0):

$$\frac{126.0}{18.0} = 7$$

So $n = 7$.

Worked example

An oxide of copper contains 88.8% copper by mass. Find the formula of the compound.

Answer

Element	Copper	Oxygen
Mass (g)	88.8	11.2
A_r	63.5	16.0
Moles	88.8/63.5 = 1.398	11.2/16.0 = 0.700
Ratio	2	1
Formula	Cu_2O	

Table 8

We can see from the data in Table 8 that the formula of the compound is Cu_2O.

Exam tip

When data are rounded to a certain number of significant figures, the moles may be not in an exact ratio but very close. Don't round the ratios too much as 1.32 is most likely 1.33, so all ratios should be multiplied by 3 to achieve whole numbers.

Balanced symbol equations and associated calculations

Balanced equations

A balanced symbol equation shows the rearrangement of atoms in a chemical reaction. Often you will be asked to write equations for reactions that are familiar and also for reactions that might be unfamiliar.

Worked example

Ammonium carbonate decomposes on heating to produce ammonia, carbon dioxide and water. Write an equation for this reaction.

Answer

Ammonium carbonate is the compound of the ammonium ion (NH_4^+) and the carbonate ion (CO_3^{2-}). So ammonium carbonate is $(NH_4)_2CO_3$.

   Balanced equation: $(NH_4)_2CO_3 \rightarrow 2NH_3 + CO_2 + H_2O$

Two NH_3 are required to balance the atoms on either side. Remember, you can never change a formula to balance an equation.

Exam tip

If given percentage data, assume you have 100g and then the percentages become masses, for example, 88.8% Cu means 88.8g of Cu. As the only two elements present are copper and oxygen, the percentage of oxygen is 100 – 88.8 = 11.2%. This means 11.2g of oxygen per 100g.

Knowledge check 13

Write an equation for the reaction of magnesium nitride with water to form magnesium hydroxide and ammonia gas.

Ionic equations

Ionic equations will be encountered in oxidation, reduction and redox reactions and also in inorganic chemistry in the student guide covering inorganic and organic chemistry 1 in this series.

An ionic equation is an equation in which the spectator ions (ions that do not take part in the reaction) are removed and only the ions that take part in the reaction are included. The charges in any ionic equation should be equal on both sides.

Worked example

Write the ionic equation for the reaction:

$$AgNO_3(aq) + NaCl(aq) \rightarrow AgCl(s) + NaNO_3(aq)$$

Answer

A white precipitate of silver(I) chloride is observed, and is shown as a solid in the written equation above. The Na^+ and NO_3^- ions remain in solution and do not take part in the formation of the white silver(I) chloride.

The ionic equation is:

$$Ag^+(aq) + Cl^-(aq) \rightarrow AgCl(s)$$

A balanced symbol equation is the key to calculations. The following steps are followed:

- Using the mass of one of the reactants, which should be given to you, calculate the amount, in moles, of this substance.
- Using the balancing numbers in the equation, calculate the amount, in moles, of whatever substance you are asked to calculate.
- Change the amount, in moles, of this substance to mass (or volume if required).

Several expressions are required to help you to do this:

Expression 1: amount, in moles $= \dfrac{\text{mass (g)}}{M_r}$

Expression 2: mass (g) = amount, in moles, $\times M_r$

Mass is often written more simply as m. Amount, in moles, is often written as n. Hence the above equations can be learned in a simplified form as long as you remember what the abbreviations are:

Expression 1: $n = \dfrac{m}{M_r}$

Expression 2: $m = n \times M_r$

Worked example

1.28 kg of magnesium nitrate are heated to constant mass. Calculate the mass of magnesium oxide formed in this reaction. Give your answer, in kg, to 3 significant figures.

Answer

$$2Mg(NO_3)_2(s) \rightarrow 2MgO(s) + 4NO_2(g) + O_2(g)$$

1.28 kg of magnesium nitrate is the information given, so, using expression 1, this can be converted to moles. However the mass must be used in grams; 1.28 kg = 1280 g

$$Mg(NO_3)_2 \quad M_r = 148.3; \quad n = \frac{m}{M_r}$$

$$n = \frac{1280}{148.3} = 8.63 \, mol$$

In the balanced symbol equation given there are balancing numbers, so this means that 2 moles of $Mg(NO_3)_2$ gives 2 moles of MgO, 4 moles of NO_2 and 1 mole of O_2.

So, 8.63 moles of $Mg(NO_3)_2$ gives 8.63 moles of MgO, 17.26 moles of NO_2 and 4.32 moles of O_2.

To calculate the mass of MgO formed in this reaction, use expression 2.

$$MgO \qquad M_r = 40.3; \qquad m = n \times M_r$$

$$m = 8.63 \times 40.3 = 347.8 \, g$$

0.348 kg of MgO are formed.

In the worked example 17.26 mol of the gas NO_2 are formed.

To find the volume of the gas at room temperature and pressure, multiply the amount, in moles, by 24 dm^3 (or 24 000 cm^3).

$$\text{volume of } NO_2 = 17.26 \times 24 = 414 \, dm^3$$

If the conditions are not room temperature and pressure, $pV = nRT$ may be used to calculated the volume, in m^3, which can be converted to dm^3 or cm^3.

Knowledge check 14

Calculate the mass of silver formed when 1.42 g of silver(I) nitrate is heated to constant mass. Give your answer to 3 significant figures.

$$2AgNO_3 \rightarrow 2Ag + O_2 + NO_2$$

Atom economy

Atom economy is a measure of how efficiently the atoms in the reactants are used in a chemical reaction.

It can be calculated as a percentage, using the expression:

$$\text{percentage atom economy} = \frac{M_r \text{ of desired product}}{\text{sum of } M_r \text{ of all reactants}} \times 100$$

Worked example

The following reaction occurs in the production of hydrogen by steam reformation of natural gas:

$$CH_4(g) + H_2O(g) \rightarrow CO(g) + 3H_2(g)$$

Calculate a value for the percentage atom economy. Give your answer to 3 significant figures.

Answer

M_r of desired product $3H_2(g) = 6.0$

sum of M_r of all reactants = $CH_4(g) + H_2O(g) = 34.0$

percentage atom economy $= \dfrac{6.0}{34.0} \times 100 = 17.6\%$

Percentage yield

Not all chemical reactions go to completion. Side reactions may also occur and there may be loss of product due to mechanical transfer between containers or loss of a volatile compound by evaporation.

The maximum mass of a product that can be obtained in a chemical reaction is called the theoretical yield. This is often not obtained and the percentage of the product obtained is called the percentage yield.

The actual mass of the product obtained is called the actual yield.

The percentage yield is calculated as:

$$\text{percentage yield} = \frac{\text{actual yield}}{\text{theoretical yield}} \times 100$$

So:

$$\text{theoretical yield} = \frac{\text{actual yield}}{\text{percentage yield}} \times 100$$

and

$$\text{actual yield} = \frac{\text{percentage yield} \times \text{theoretical yield}}{100}$$

Exam tip

The yields (actual and theoretical) may be mass values (which is more usual) or can be number of moles. Either is correct as long as theoretical and actual are both given in the same units.

Worked example 1

Calcium carbide, CaC_2, is manufactured from lime, CaO, and coke, C, in an electric arc furnace.

$$CaO(s) + 3C(s) \rightarrow CaC_2(s) + CO(g)$$

Calculate the percentage yield of calcium carbide if 245 g is manufactured from 300 g of lime assuming that the coke is in excess.

Answer

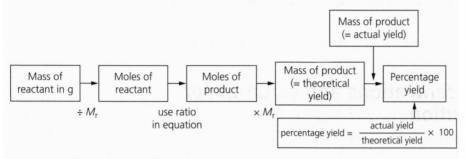

mass of reactant (CaO) = 300 g

moles of reactant (CaO) = $\dfrac{300}{56.1}$ = 5.35   (M_r of CaO = 56.1)

moles of product (CaC_2) = 5.35   (1 : 1 ratio in equation)

theoretical yield of product = 5.35 × 64.1 = 342.9 g   (M_r of CaC_2 = 64.1)

percentage yield = $\dfrac{245}{342.9}$ × 100 = 71.4%

Worked example 2

12.8 g of magnesium nitrate were obtained from the reaction between magnesium and nitric acid, assuming a 77.0% yield. Calculate the mass of magnesium required assuming an excess of nitric acid.

Answer

$$Mg + 2HNO_3 \rightarrow Mg(NO_3)_2 + H_2$$

actual yield of magnesium nitrate = 12.8 g

percentage yield = $\dfrac{\text{actual yield}}{\text{theoretical yield}}$ × 100

so, theoretical yield = $\dfrac{\text{actual yield}}{\text{percentage yield}}$ × 100

theoretical yield of magnesium nitrate (i.e. 100% yield) = $\dfrac{12.8}{77.0}$ × 100 = 16.6 g

theoretical yield (in moles) of magnesium nitrate = $\dfrac{16.6}{148.3}$ = 0.112 mol

moles of Mg required = 0.112 as 1 : 1 ratio of $Mg : Mg(NO_3)_2$

mass of Mg required = 0.112 × 24.3 = 2.72 g

Knowledge check 15

14.0 g of aluminium oxide were formed by reacting 10.0 g of aluminium with excess oxygen.

$$4Al + 3O_2 \rightarrow 2Al_2O_3$$

Calculate the percentage yield to 3 significant figures.

Understanding atom economy and percentage yield

Chemists often use percentage yield to determine the efficiency of a chemical synthesis process. A high percentage yield would indicate that the reaction process is efficient in converting reactants into products. This is important for profit, but percentage yield does not take into account any waste products.

Chemistry is like all other industries that are concerned about their effects on the environment, particularly when it comes to waste. A reaction may have a high percentage yield but have a low atom economy. This would mean that other products in the reaction will be waste and with a high percentage yield there are just more of them.

Chemists now look at having a high percentage yield but also a high atom economy to reduce waste. This is the main thrust of green chemistry.

Calculating concentrations and volumes for reactions in solution

Solutions are often reacted together in a process called a titration, which involves the use of an indicator.

Practical techniques of titration

A titration is a method of volumetric analysis. The main pieces of apparatus used in a titration are a burette, a pipette with safety filler, a volumetric flask and several conical flasks. One solution is placed in a burette and the other is placed in a conical flask. An indicator is added to the solution in the conical flask. The solution in the burette is then gradually added to the solution in the conical flask. The end point of the titration is shown when the indicator changes colour. This is the point at which the reaction is complete.

Preparing and using a burette

To use a burette in a titration:
- rinse the burette with deionised water
- ensure that the water flows through the jet
- discard the water
- rinse the burette with the solution you will be filling it with
- ensure that the solution flows through the jet
- discard the solution
- charge (fill) the burette with the solution you will be using in it
- ensure that the volume of solution the burette contains is read at the bottom of the meniscus, as shown in Figure 8

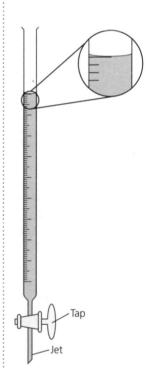

Figure 8 Burette showing the meniscus

Preparing and using a pipette

Pipettes measure out exactly $25.0\,cm^3$ or $10.0\,cm^3$. To use a pipette in a titration:
- use a pipette filler to rinse the pipette using deionised water
- discard the water
- rinse the pipette with the solution you will be filling it with
- discard this solution

- draw up the solution above the line on the pipette
- release the solution until the meniscus sits on the line
- release the solution in the pipette into a conical flask

(Remember that $1 cm^3 = 1 ml$ but chemists prefer cm^3 as their volume unit.)

An exact volume of a solution is vital in volumetric work as taking exactly $25.0 cm^3$ of a known concentration of a solution means that we know exactly how many moles of the dissolved substance are present in the conical flask.

Conical flasks

Conical flasks are used in titrations as they can be swirled easily to mix the reactants. Also, the sloped sides on conical flasks help to prevent any of the solution spilling out when it has been added. The conical flask should be rinsed with deionised water before use.

The conical flask does not have to be completely dry before use as the exact volume of solution added contains an exact number of moles of solute. Extra deionised water does not add to the number of moles of solute.

Volumetric flasks

Volumetric flasks (Figure 9) are used when diluting one of the solutions before the titration is carried out. They can also be used when preparing a solution of a solid.

Carrying out a dilution of a solution

To use a volumetric flask for a dilution you should take the following steps:
- pipette $25.0 cm^3$ of the original solution into a clean volumetric flask
- add deionised water to the flask until the water is just below the line
- using a disposable pipette add deionised water very slowly until the bottom of the meniscus is on the line
- stopper the flask and invert to mix thoroughly

Dilution factor

The dilution factor is the amount the original solution is diluted by. It is calculated by dividing the new total volume by the volume of original solution put into the mixture. For example:
- If a $25.0 cm^3$ sample of a solution is made up to a total volume of $250 cm^3$ using deionised water then the dilution factor is 10.
- If a $10.0 cm^3$ sample of a solution is made up to a total volume of $250 cm^3$ using deionised water then the dilution factor is 25.

Preparing a solution from a mass of solid

When preparing a solution from a solid it is important not to lose any of the solid or solution before it is placed in the volumetric flask. To prepare a solution from a solid you should carry out the following steps:
- weigh out an accurate mass of a solid in a weighing boat and dissolve it in a suitable volume of deionised water in a beaker. Rinse the weighing boat into the beaker with deionised water — stir with a glass rod

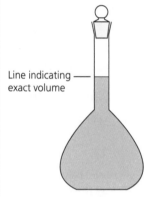

Line indicating exact volume

Figure 9 A volumetric flask

- when the solid has dissolved, hold the glass rod above the beaker and rinse it with deionised water before removing it
- place a glass funnel into the top of a clean volumetric flask and pour the prepared solution down a glass rod into the funnel
- rinse the glass rod with deionised water into the funnel
- rinse the funnel with deionised water
- remove the funnel and add deionised water to the volumetric flask until the water is just below the line
- using a disposable pipette add deionised water very slowly until the bottom of the meniscus is on the line
- stopper the flask and invert to mix thoroughly

Carrying out a titration

The major points of carrying out a titration are shown in Figure 10.

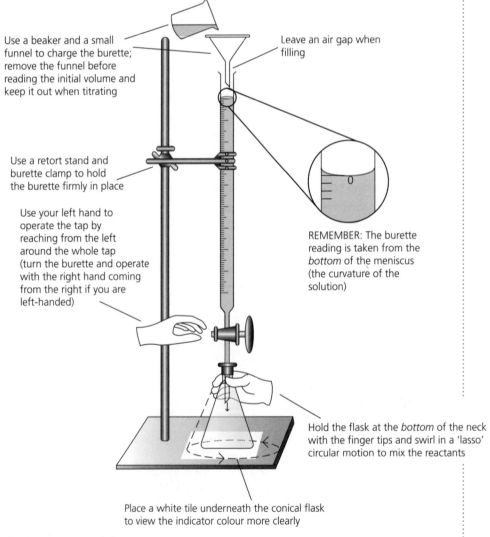

Use a beaker and a small funnel to charge the burette; remove the funnel before reading the initial volume and keep it out when titrating

Leave an air gap when filling

Use a retort stand and burette clamp to hold the burette firmly in place

Use your left hand to operate the tap by reaching from the left around the whole tap (turn the burette and operate with the right hand coming from the right if you are left-handed)

REMEMBER: The burette reading is taken from the *bottom* of the meniscus (the curvature of the solution)

Hold the flask at the *bottom* of the neck with the finger tips and swirl in a 'lasso' circular motion to mix the reactants

Place a white tile underneath the conical flask to view the indicator colour more clearly

Figure 10 Carrying out a titration

For a single titration, consecutive titrations should be carried out until concordant results are obtained. A first rough titration should be carried out and allowed to overshoot to obtain a rough idea of where the end point will be. In subsequent titrations the solution should be added drop-wise near the end point.

Standard solutions are used in volumetric analysis. A standard solution is a solution of known concentration.

A burette has a total graduated volume of $50.0\,cm^3$. You can perform two titrations using a burette if the titres are well below $25.0\,cm^3$. However, if the rough titration result is close to or above $25.0\,cm^3$, it is essential that you refill the burette before starting the first accurate titration.

Recording titration results

The titre is the volume delivered from the burette into the conical flask until the indicator changes colour (the end point) during a titration.

Note the following points about recording titration results:

- Units are stated with the row headings and are never put in the main body of the table (see Table 9).
- All values should be recorded to $0.05\,cm^3$; 0 is written as 0.00; 17.5 is written as 17.50; the second decimal place should always be given, but must be 0 or 5.
- When calculating the average titre, ignore the rough titration and any result that is clearly not within $0.10\,cm^3$ of the other accurate titration values. Titres within $0.10\,cm^3$ of each other are called concordant results and only concordant results should be used to calculate the average titre. Outliers (results which are not concordant) are ignored.

Table 9 shows a sample titration results table.

	Rough	1	2	3
Final burette reading/cm³	24.50	24.00	24.00	23.95
Initial burette reading/cm³	0.00	0.00	48.40	0.00
Titre/cm³	24.50	24.00	24.40	23.95

Table 9

The average titre from the results shown in Table 9 is calculated using titrations 1 and 3 as these are the concordant results. The average titre is $23.98\,cm^3$. This is the average of $24.00\,cm^3$ and $23.95\,cm^3$ and the result is given to 2 decimal places. The rough result and titration 2 are ignored.

Common acids and bases

Strong acids: HCl, H_2SO_4, HNO_3 Weak acids: CH_3COOH; organic acids

Strong bases: $NaOH$, KOH Weak bases: NH_3, Na_2CO_3

Common equations used in titrations:

$NaOH + HCl \rightarrow NaCl + H_2O$ 1 : 1 ratio for NaOH : HCl

Exam tip

Units of volume are in cm^3 ($1\,cm^3 = 1\,ml$). Concentration units are $mol\,dm^{-3}$. Remember that $1\,dm^3$ is the same as 1 litre. Units of $g\,dm^{-3}$ (grams per dm^3) may be calculated by multiplying the concentration (in $mol\,dm^{-3}$) by the M_r of the solute.

Knowledge check 16

State two common strong acids.

$$2NaOH + H_2SO_4 \rightarrow Na_2SO_4 + 2H_2O \qquad \text{2:1 ratio for } NaOH:H_2SO_4$$

$$NaOH + CH_3COOH \rightarrow CH_3COONa + H_2O \qquad \text{1:1 ratio for } NaOH:CH_3COOH$$

All of the above can be rewritten with KOH and the same ratio applies.

$$Na_2CO_3 + 2HCl \rightarrow 2NaCl + CO_2 + H_2O \qquad \text{1:2 ratio for } Na_2CO_3:HCl$$

$$Na_2CO_3 + H_2SO_4 \rightarrow Na_2SO_4 + CO_2 + H_2O \qquad \text{1:1 ratio for } Na_2CO_3:H_2SO_4$$

The ratio of the solutes is important as it allows a calculation of the ratio between the number of moles of the solution added from the burette and the number of moles of the other substance in the conical flask.

Indicators

The two main indicators used for acid–base titrations are phenolphthalein and methyl orange (Table 10).

	Indicator	
	Phenolphthalein	Methyl orange
Colour in acidic solutions	Colourless	Red
Colour in neutral solutions	Colourless	Orange
Colour in alkaline solutions	Pink	Yellow
Titrations suitable for:	Strong acid–strong base Weak acid–strong base	Strong acid–strong base Strong acid–weak base

Table 10

You need to know the colour change of methyl orange and phenolphthalein (Table 11).

Titration	Methyl orange	Phenolphthalein
Acid in conical flask/alkali in burette	Red to yellow	Colourless to pink
Alkali in conical flask/acid in burette	Yellow to red	Pink to colourless

Table 11

> **Exam tip**
>
> Make sure that you know what solution is present in the conical flask and what solution is being added from the burette to determine the colour change. For example, if sodium hydroxide is being added from the burette to hydrochloric acid in the conical flask, with phenolphthalein as the indicator, then the colour change is colourless to pink.

Typical method of acid–base titration

- The solution of known concentration is usually placed in the burette.
- Using a pipette filler rinse the pipette with deionised water and with the solution you are going to pipette into the conical flask. Pipette 25.0 cm³ of this solution into three different conical flasks.
- Add three to five drops of suitable indicator to each conical flask.

> **Exam tip**
>
> The choice of indicator is based on the type of acid and base in the titration. It is important to be able to choose the correct indicator for a particular titration. If ethanoic acid (a weak acid) and sodium hydroxide solution (a strong alkali) are being used then phenolphthalein is the indicator of choice. Both methyl orange and phenolphthalein can be used for strong acid–strong base titrations. This is mainly examined at A-level.

> **Knowledge check 17**
>
> State the colour change when hydrochloric acid is added to sodium hydroxide solution containing phenolphthalein.

- Note the colour of the indicator in this solution.
- Rinse the burette with deionised water and with the solution you are using to fill it. Fill burette with this solution. Titrate until indicator just changes colour, adding drop-wise near the end point.
- Repeat for accuracy and calculate the average titre from titre values that are within $0.10\,cm^3$ of each other.

Volumetric calculations

As we have already seen, the amount, in moles, of a substance may be calculated from the volume, in cm^3, and concentration, in $mol\,dm^{-3}$, of the solution.

The expression used is:

$$\text{amount, in moles} = \frac{\text{volume }(cm^3) \times \text{concentration }(mol\,dm^{-3})}{1000}$$

$$\text{or } n = \frac{v \times c}{1000}$$

Typical acid–base calculation question

A solution of oven cleaner contains sodium hydroxide. $25.0\,cm^3$ of this solution were pipetted into a $250\,cm^3$ volumetric flask and the solution made up to the mark with deionised water. $25.0\,cm^3$ of this solution were placed in a conical flask and titrated against $0.0215\,mol\,dm^{-3}$ hydrochloric acid using phenolphthalein indicator. The average titre was determined to be $17.20\,cm^3$. Calculate the concentration of the sodium hydroxide solution in $mol\,dm^{-3}$. Give your answer to 3 decimal places.

The answer can be laid out clearly as follows:

$$\text{moles of HCl} = \frac{v \times c}{1000} = \frac{17.20 \times 0.0215}{1000} = 3.698 \times 10^{-4}\,mol$$

So $3.698 \times 10^{-4}\,mol$ of HCl were added from the burette to reach the end-point.

The equation for the reaction occurring in the conical flask is:

$$NaOH + HCl \rightarrow NaCl + H_2O$$

The ratio of NaOH : HCl is 1 : 1 so the amount, in moles, of NaOH in $25.0\,cm^3$ in the conical flask = $3.698 \times 10^{-4}\,mol$

The concentration of the diluted solution may now be calculated by rearranging the expression:

$$n = \frac{v \times c}{1000}$$

$$\text{concentration of diluted solution} = \frac{n \times 1000}{v}$$

$$\frac{3.698 \times 10^{-4} \times 1000}{25.0} = 0.0148\,mol\,dm^{-3}$$

dilution factor = 10 (as $25.0\,cm^3$ is diluted to $250\,cm^3$)

concentration of undiluted solution = $0.0148 \times 10 = 0.148\,mol\,dm^{-3}$

The undiluted oven cleaner solution of sodium hydroxide has a concentration of $0.148\,mol\,dm^{-3}$.

Exam tip

These questions are often structured and follow a pattern — the first calculation will be an amount, in moles, of a solute from a solution volume and concentration. As the solution is diluted the terms 'undiluted solution' and the 'diluted solution' are often used.

An alternative method of carrying out the calculation from the moles of NaOH in the conical flask is as follows:

Amount, in moles, of NaOH in 250 cm^3 volumetric flask = $3.698 \times 10^{-4} \times 10$
= 3.698×10^{-3} mol

(as 25.0 cm^3 were taken out of a total volume of 250 cm^3 so a tenth of the moles of solute were taken out).

The total moles of NaOH in the volumetric flask came from the 25.0 cm^3 of the undiluted oven cleaner solution, so the moles of NaOH in 25.0 cm^3 of the undiluted solution = 3.698×10^{-3} mol.

$$\text{concentration} = \frac{n \times 1000}{v} = \frac{3.698 \times 10^{-3} \times 1000}{25.0} = 0.148 \, \text{mol dm}^{-3}$$

This style of an acid–base titration is common and the only difference may be the ratio in the reaction and whether or not there is a dilution.

Degree of hydration titrations

Titrations can also be used to determine the degree of hydration of a salt, but the salt in solution must react with an acid, so although hydrated sodium carbonate is a common example, it could be applied to any hydrated carbonate.

Typical degree of hydration titration

- Dissolve a known mass of the hydrated salt (usually hydrated sodium carbonate) in water and make up volume to 250 cm^3.
- Pipette 25.0 cm^3 of the solution into conical flask.
- Titrate with hydrochloric acid of known concentration using methyl orange indicator.
- Average titre can be used to calculate the moles of hydrochloric acid.

  $Na_2CO_3 + 2HCl \rightarrow 2NaCl + CO_2 + H_2O$
- Calculate moles of sodium carbonate in 25.0 cm^3 of solution using the equation ratio above.
- Calculate the moles of sodium carbonate in 250.0 cm^3 of solution.
- Use mass and moles in 250.0 cm^3 to calculate M_r of $Na_2CO_3.xH_2O$.
- Subtract 106.0 for Na_2CO_3 — this will give you M_r of all the water.
- Divide by 18.0 to calculate moles of water.

Typical degree of hydration calculation question

1.96 g of hydrated sodium carbonate, $Na_2CO_3.xH_2O$, are dissolved in deionised water, placed in a volumetric flask and the volume made up to 250.0 cm^3 using deionised water. A 25.0 cm^3 sample of this solution was pipetted into a conical flask and titrated against 0.140 mol dm^{-3} hydrochloric acid using methyl orange indicator. The average titre was found to be 11.20 cm^3. Calculate a value for x in $Na_2CO_3.xH_2O$.

$Na_2CO_3 + 2HCl \rightarrow 2NaCl + H_2O + CO_2$

Lay out your answer clearly:

$$\text{amount, in moles, of HCl used} = \frac{v \times c}{1000} = \frac{11.20 \times 0.140}{1000} = 1.568 \times 10^{-3} \, \text{mol}$$

Exam tip

Concentration may also be measured in g dm^{-3}. To convert between mol dm^{-3} and g dm^{-3}, multiply by the M_r of the solute. The M_r of NaOH is 40.0.

$0.148 \times 40.0 = 5.92$ g dm^{-3}

Knowledge check 18

Calculate the concentration of hydrochloric acid if 25.0 cm^3 is completely neutralised by 17.25 cm^3 of 0.143 mol dm^{-3} sodium hydroxide solution. Give your answer to 3 significant figures.

Ratio of $Na_2CO_3 : HCl$ is $1:2$,

so moles of Na_2CO_3 in $25.0\,cm^3 = \dfrac{1.568 \times 10^{-3}}{2}$

$$= 7.84 \times 10^{-4}\,mol$$

amount, in moles, of Na_2CO_3 present in $250\,cm^3 = 7.84 \times 10^{-4} \times 10 = 7.84 \times 10^{-3}\,mol$

The amount of moles of Na_2CO_3 present in solution is the same as the number of moles of solid $Na_2CO_3.xH_2O$ added.

amount, in moles, of $Na_2CO_3.xH_2O = 7.84 \times 10^{-3}\,mol$

M_r of $Na_2CO_3.xH_2O = \dfrac{mass}{moles} = \dfrac{1.96}{7.84 \times 10^{-3}} = 250$

M_r of $Na_2CO_3 = 106.0$ so mass due to water of crystallisation $= 250 - 106 = 144$.

value of $x = \dfrac{144}{18} = 8$

In this sample of $Na_2CO_3.xH_2O$, $x = 8$.

Back titration

A back titration is used to determine the purity of a group 2 metal, group 2 oxide or group 2 carbonate. It is used when the substance under analysis is not soluble in water but will react with an acid. A known mass of the insoluble solid is added to excess acid and the acid that is left over is titrated with a standard solution of an alkali such as sodium hydroxide solution.

Typical method of back titration

- A known mass of an insoluble metal, metal oxide or metal carbonate is added to an excess of a known volume and concentration of acid (usually hydrochloric acid).
- Once the reaction is complete the mixture may be diluted to $250\,cm^3$ in a volumetric flask.
- Titrate $25.0\,cm^3$ of this solution (containing the excess acid) against sodium hydroxide solution of known concentration using phenolphthalein indicator.

Typical back titration calculation question

Two indigestion tablets containing calcium carbonate are placed in $25.0\,cm^3$ of $2.00\,mol\,dm^{-3}$ hydrochloric acid. Once the reaction was complete the solution was transferred to a volumetric flask and the volume made up to $250\,cm^3$ using deionised water. $25.0\,cm^3$ of this solution was titrated against $0.158\,mol\,dm^{-3}$ sodium hydroxide solution and the average titre was determined to be $22.90\,cm^3$. Calculate the mass of calcium carbonate, in mg, present in one indigestion tablet.

$CaCO_3(s) + 2HCl(aq) \rightarrow CaCl_2(aq) + CO_2(g) + H_2O(l)$

$NaOH(aq) + HCl(aq) \rightarrow NaCl(aq) + H_2O(l)$

amount, in moles, of NaOH which reacted $= \dfrac{v \times c}{1000} = \dfrac{22.90 \times 0.158}{1000} = 3.618 \times 10^{-3}\,mol$

amount, in moles, of HCl in $25.0\,cm^3$ in the conical flask $= 3.618 \times 10^{-3}\,mol$ (1:1 ratio)

amount, in moles, of HCl left over from the reaction with $CaCO_3 = 3.618 \times 10^{-3} \times 10$

$= 0.03168\,mol$ (all that was left over was present in $250\,cm^3$)

initial amount, in moles, of HCl added $= \dfrac{v \times c}{1000} = \dfrac{25.0 \times 2.00}{1000} = 0.0500\,mol$

amount, in moles, of HCl that reacted with $CaCO_3 = 0.0500 - 0.03168 = 0.01382\,mol$

amount, in moles of $CaCO_3 = \dfrac{0.01382}{2} = 6.91 \times 10^{-3}\,mol$

mass of $CaCO_3 = 6.91 \times 10^{-3} \times 100.1 = 0.692\,g$ (M_r of $CaCO_3 = 100.1$)

mass of $CaCO_3$, in mg, $= 0.692 \times 1000 = 692\,mg$

mass of $CaCO_3$ in one indigestion tablet $= \dfrac{692}{2} = 346\,mg$

Knowledge check 20

Calculate the percentage purity of a sample of magnesium metal if 1.00 g of magnesium is reacted with 25 cm^3 of 1.5 mol dm^{-3} hydrochloric acid. The excess acid requires 14.2 cm^3 of 0.745 mol dm^{-3} sodium hydroxide solution.

Required practical 1

Make up a volumetric solution and carry out a simple acid–base titration

This practical may ask you to determine the concentration of ethanoic acid in vinegar, or the mass of calcium carbonate in an indigestion tablet, or the M_r of a metal carbonate. To carry out this practical you must be familiar with the content on pages 28 to 36.

Exam tip

Using a lower concentration of acid in a back titration would give a larger titre so a larger percentage error in the titration.

Exam tip

For titres taken from two volume measurements of a burette and temperature changes taken from two temperature readings on a thermometer, the error is multiplied by 2 before calculating the percentage error.

Percentage error

The percentage error in measurements obtained from different pieces of apparatus can be calculated from the error on the instrument and the value of the quantity being measured. For example:

- a volume of 25 cm^3 is measured using a measuring cylinder that has an error of ± 0.5 cm^3. The percentage error is $\dfrac{0.5}{25} \times 100 = 2\%$.
- a titre of 24.50 cm^3 is determined from two volume readings on a burette, where the burette has an error of ± 0.05 cm^3. As two volume measurements are used to calculate the titre, the total error is ± 0.1 cm^3. The percentage error is $\dfrac{0.1}{24.50} \times 100 = 0.41\%$.

Summary

- A balanced symbol equation is the ratio of the number of moles of each reactant and product.
- 1 mole of a substance is the mass in grams that contains the Avogadro constant of particles (6.02×10^{23}).
- An empirical formula is the simplest ratio of the atoms in a compound.
- Empirical formulae are used for giant covalent compounds and ionic compounds.
- Heating a hydrated compound to constant mass produces an anhydrous compound.

- The simplest ratio of the number of moles of the anhydrous compound to the number of moles of water determines the degree of hydration.
- percentage yield $= \dfrac{\text{actual yield}}{\text{theoretical yield}} \times 100$
- atom economy $= \dfrac{M_r \text{ of desired product}}{\text{sum of } M_r \text{ of all reactants}} \times 100$

- A titration is carried out using a pipette and burette and two solutions are mixed to determine the exact volume of one solution required to react with an exact volume of the other solution.
- Acid-base titrations in which one solution is an acid and the other is a base (alkali) are one of the most common types of titrations.
- An indicator is used to determine the exact point when the acid has neutralised the base or vice versa.
- Two common indicators are phenolphthalein and methyl orange.
- Preparation (including rinsings) and accurate use of the apparatus are important for obtaining reliable results.
- The volume of solution added from the burette is called the titre. One rough titration and subsequent titrations are carried out until two concordant titres that are within 0.1 cm^3 of each other are obtained.
- Based on the average titre, calculations are carried out to determine concentration, number of moles, mass, M_r, percentage purity, degree of hydration and even to identify unknown elements.

Bonding

Ionic bonding

Ionic bonding is the electrostatic attraction between oppositely charged ions in a regular, ionic lattice. A lattice is a regular arrangement of particles, in this case positive and negative ions.

Ionic compounds are formed when electrons are transferred, usually from metal atoms to non-metal atoms, forming ions. A positive ion (cation) is formed when electrons are lost and a negative ion (anion) is formed when electrons are gained. Ions (apart from d block elements) have noble gas electron configuration and are stable.

Formation of ions

The following examples show how atoms lose or gain electrons to form ions:
- In sodium chloride, the sodium atom loses its $3s^1$ electron to form a sodium ion, Na^+, with a noble gas electron configuration. The chlorine atom gains one electron to form a chloride ion, Cl^-, also with a noble gas electron configuration.

Sodium atom | Sodium ion

Na $\rightarrow$ Na$^+$

$1s^2\, 2s^2\, 2p^6\, (3s^1)$ | $1s^2\, 2s^2\, 2p^6$

Chlorine atom | Chloride ion

Cl $\rightarrow$ Cl$^-$

$1s^2\, 2s^2\, 2p^6\, 3s^2\, 3p^5$ | $1s^2\, 2s^2\, 2p^6\, 3s^2\, 3p^6$

- In calcium fluoride, two fluorine atoms are required for each calcium atom as each calcium atom has two electrons to lose from $4s^2$ to obtain a noble gas configuration. Each fluorine atom requires one electron to achieve a noble gas configuration.

> **Exam tip**
> You may often be asked to predict the type of bonding present in a substance. Generally, compounds containing a metal, particularly a group 1 or 2 metal, and a non-metal, particularly a group 6 or 7 non-metal, have ionic bonding.

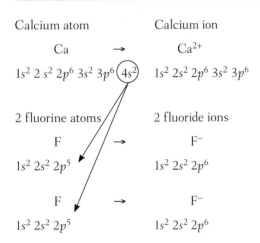

Calcium atom

$$Ca \longrightarrow$$

$$1s^2\ 2\ s^2\ 2p^6\ 3s^2\ 3p^6\ \boxed{4s^2}$$

Calcium ion

$$Ca^{2+}$$

$$1s^2\ 2s^2\ 2p^6\ 3s^2\ 3p^6$$

2 fluorine atoms

$$F \longrightarrow$$

$$1s^2\ 2s^2\ 2p^5$$

2 fluoride ions

$$F^-$$

$$1s^2\ 2s^2\ 2p^6$$

$$F \longrightarrow$$

$$1s^2\ 2s^2\ 2p^5$$

$$F^-$$

$$1s^2\ 2s^2\ 2p^6$$

Exam tip

The attraction between the positive and negative ions is the ionic bond. It is not the transfer of electrons. This is a common error.

Nature of covalent and dative covalent bonds

The covalent bond

A single covalent bond contains a shared pair of electrons. Normally each atom provides one electron. A single covalent bond is represented as a line between two atoms, for example, H–Cl.

Multiple bonds contain multiple shared pairs of electrons.

A double covalent bond is two shared pairs of electrons. A double covalent bond is represented as double line between two atoms, for example, O=C=O.

A triple covalent bond is three shared pairs of electrons. A triple covalent bond is represented as a triple line between two atoms, for example, N≡N.

Covalent bonds exist between non-metal atoms (some exceptions occur where metal atoms can form covalent bonds, for example, Be in $BeCl_2$ and Al in $AlCl_3$). A shared pair of electrons can be represented as ×● to show that the two electrons in the bond are from different atoms (Figure 11). A bonding pair of electrons is a pair of electrons shared between two atoms. A lone pair of electrons is an unshared (non-bonding) pair of electrons in the outer shell of an atom. It can be represented as ●● or ×× or ●● or ×× or ⌒.

Knowledge check 21

State how a covalent bond is formed between carbon and hydrogen.

H×C̈l̈:

Lone pair

H–Cl

H:N̈:H
  |
  H

Bonding pair

H–N–H
   |
   H

:Ö:C:Ö:

Two bonding pairs
= double bond

O=C=O

×N:::N:— lone pair

Three bonding pairs
= triple bond

N≡N

Figure 11 Dot and cross diagrams

The coordinate bond (dative covalent)

A **coordinate** (dative covalent) bond contains a shared pair of electrons with both electrons supplied by one atom. Note that when a coordinate bond forms, a lone pair

Exam tip

Covalent bonds are found in compound ions, for example, NH_4^+, NO_3^-, SO_4^{2-}, CO_3^{2-} and HCO_3^-, and in giant covalent and molecular structures (see pages 51 and 52).

of electrons becomes a bonding pair of electrons. Once formed, the coordinate bond is identical to a normal covalent bond.

Example: the ammonium ion, NH_4^+

When an ammonia (NH_3) molecule reacts with an H^+ ion, the ammonia molecule donates its lone pair of electrons to the H^+. Both electrons in the coordinate bond are supplied by the one atom (N). The coordinate bond is shown as an arrow in a bonding diagram, with the direction indicating where the two electons come from.

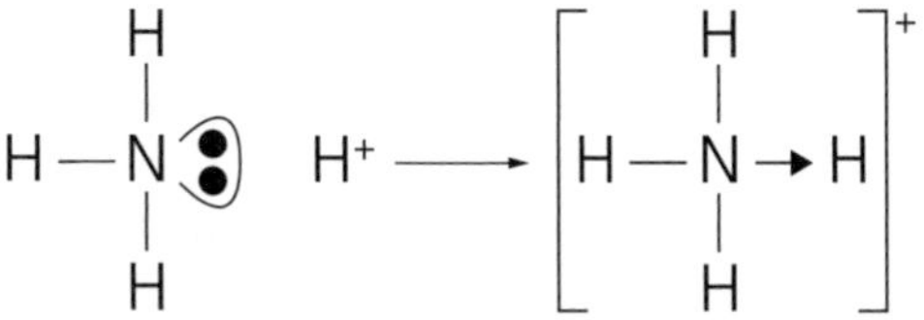

$$H-N\!\!: \quad H^+ \longrightarrow \left[H-N \rightarrow H \right]^+$$

Figure 12 The formation of NH_4^+

Example: BF_4^- formed from BF_3 and F^-

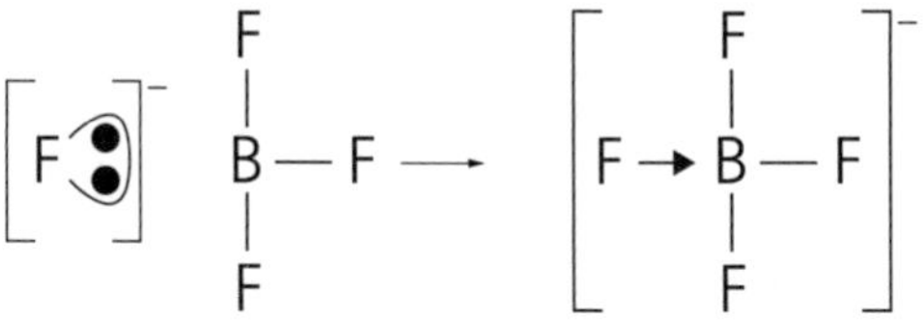

$$\left[F\!\!: \right]^- \quad B-F \longrightarrow \left[F \rightarrow B-F \right]^-$$

Figure 13 The formation of BF_4^-

Metallic bonding

A metallic bond is the attraction between delocalised electrons and positive ions arranged in a lattice. Delocalised electrons are those that are not confined to any one atom and can move throughout the lattice.

Magnesium

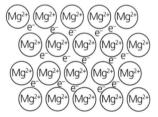

Figure 14 Metallic lattice of magnesium

Shapes of simple molecules and ions

Molecules and ions are three dimensional and have different shapes.

- The shape of a molecule or ion depends on the number of pairs of electrons in the outer shell of the central atom.
- Bonding pairs and lone pairs of electrons exist as charge clouds that repel each other.
- Electron pairs maximise their distance apart to minimise repulsion between them.
- Lone pair–lone pair repulsion is greater than lone pair–bond pair repulsion, which is greater than bond pair–bond pair repulsion. This is known as the valence shell

electron pair repulsion theory (VSEPR). As a result of these differences in size of repulsion, the greatest angles in the shape of the molecule will be between lone pairs of electrons. Bond angles between bonded pairs are often reduced because they are pushed together by lone pair repulsion.

Calculating the number of electron pairs

To work out the shape of a molecule or ion you need to know the number and type of electron pairs on the central atom (the one that the other atoms are bonded to). Use the following method.

1 Use your periodic table to find the number of outer shell electrons around the central atom.

2 Add one if it is a negative ion, subtract one if it is a positive ion.

3 The formula will tell you how many atoms are bonded to the central atom — each will share one of their atoms — add one for each atom shared.

4 You now have the total number of electrons around the central atom — divide this number by two to find the total number of electron pairs.

5 Take away the number of bonded atoms to find the number of lone pairs.

Example: CF_4

■ Carbon, the central atom has four electrons in its outer shell.
■ There are four fluorine atoms bonded to carbon, so there are four electrons coming from the fluorine atoms, giving a total of eight electrons
■ Eight electrons = four electron pairs.
■ Four electron pairs – four bonded atoms = no lone pairs.
■ In CF_4 there are four bonding pairs.

Example: BrF_4^-

■ Bromine has seven electrons in its outer shell.
■ There is a 1– charge so add one to give eight electrons.
■ There are four fluorine atoms bonded, so there are four electrons coming from the fluorine atoms, giving a total of twelve electrons.
■ Twelve electrons = six electron pairs.
■ Six electron pairs – four bonded atoms = two lone pairs.
■ In BrF_4^- there are four bonding pairs and two lone pairs.

Stating and explaining the shape of a molecule

1 State the number of lone pairs of electrons and bonding pairs of electrons. (Work this out using the method above.)

2 State that the 'electron pairs repel each other'.

3 a If only bonding pairs of electrons are present, state that bonding pairs of electrons repel each other equally, *or*

   b If both bonding pairs of electrons and lone pairs of electrons are present, state that lone pairs of electrons repel more than bonding pairs of electrons.

4 State that the molecule takes up the shape to minimise repulsions.

5 State the shape and the bond angle.

Exam tip

In questions on this topic you may be asked for any combination of: a sketch of the shape; the name of the shape; the bond angle; an explanation of the shape, for molecules and ions with up to six electron pairs surrounding the central atom.

Examples with only bonding pairs of electrons

Central atoms with two electron pairs, for example beryllium chloride, $BeCl_2$

Figure 15 shows that $BeCl_2$ has a linear shape, with a bond angle of 180°.

Explanation of shape:
- electron pairs repel each other
- two bonding pairs of electrons repel each other equally
- molecule takes a **linear** shape to minimise repulsions

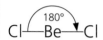

Figure 15 The shape of $BeCl_2$

> **Exam tip**
>
> For each of these examples ensure that you can work out the number and type of electron pairs using the method above. For example, in PCl_5 there are five outer electrons in P plus five from the Cl atoms = ten electrons = five pairs of electrons. Taking away the five bonding pairs of electrons indicates that there are no lone pairs in this molecule.

Central atoms with three electron pairs, for example boron trifluoride, BF_3

Figure 16 shows that BF_3 has a trigonal planar shape, with a bond angle of 120°.

Explanation of shape:
- electron pairs repel each other
- three bonding pairs of electrons repel each other equally
- molecule takes a **trigonal planar** shape to minimise repulsions

Figure 16 The shape of BF_3

> **Exam tip**
>
> When drawing a three-dimensional shape, for example, tetrahedral, bonds in the plane of the page are shown as normal lines (—). Bonds coming out of the plane of the page are drawn using a solid wedge ◢. Bonds going backwards from the plane of the paper are shown using a dashed line (– – –).

Central atoms with four electron pairs, for example methane, CH_4

Figure 17 shows that CH_4 has a tetrahedral shape, with a bond angle of 109.5°.

Explanation of shape:
- electron pairs repel each other
- four bonding pairs of electrons repel each other equally
- molecule takes a **tetrahedral** shape to minimise repulsions

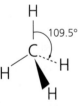

Figure 17 The shape of CH_4

Central atoms with five electron pairs, for example phosphorus pentachloride, PF_5

Figure 18 shows that the shape of PF_5 is trigonal bipyramidal, with bond angles of 120° and 90°.

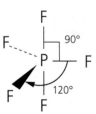

Figure 18 The shape of PF_5

Explanation of shape:

- electron pairs repel each other
- five bonding pairs of electrons repel each other equally
- molecule takes a **trigonal bipyramidal** shape to minimise repulsions

Central atoms with six electron pairs, for example sulfur hexafluoride, SF$_6$

Figure 19 shows that the shape of SF$_6$ is octahedral, with bond angles of 90°.

Explanation of shape:

- electron pairs repel each other
- six bonding pairs of electrons repel each other equally
- molecule takes up an **octahedral** shape to minimise repulsions

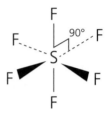

Figure 19 The shape of SF$_6$

> **Exam tip**
>
> SF$_6$ has an expanded octet — there are 12 electrons around the central sulfur atom. This can happen with atoms in groups 5, 6 and 7.

Examples with bonding pairs of electrons and lone pairs of electrons

Remember that with any shape with lone pairs of electrons, you should show the lone pairs on the central atom. You can do this using •• above the atom or even ××. Sometimes the orbital containing the lone pair of electrons is shown as ◯ above the atom. For example, all the diagrams in Figure 20 would be accepted for a sketch of the shape of an ammonia molecule.

Ammonia, NH$_3$

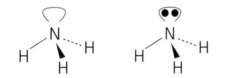

Figure 20 Different ways to show a lone pair of electrons

Figure 21 shows that NH$_3$ is pyramidal, with a bond angle of 107°.

Explanation of shape:

- electron pairs repel each other
- three bonding pairs of electrons and one lone pair of electrons
- lone pair of electrons has a greater repulsion
- molecule takes a **pyramidal** shape to minimise repulsions

Figure 21 The shape of NH$_3$

> **Exam tip**
>
> The basic arrangement of the electron pairs is tetrahedral around the nitrogen, but as there is no atom attached to the lone pair all you see is the bottom of the tetrahedron, which looks like a pyramid. The extra repulsion from the lone pair squeezes the bonds closer, decreasing the bond angle to 107°.

Water, H₂O

Figure 22 shows that H_2O has a bent shape, with a bond angle of 104.5°.

Explanation of shape:

- electron pairs repel each other
- two bonding pairs of electrons and two lone pairs of electrons
- lone pairs of electrons have a greater repulsion
- molecule takes a **bent** shape to minimise repulsions

Figure 22 The shape of H_2O

Examples involving coordinate bonds

When a coordinate bond forms it converts a lone pair of electrons into a bonding pair of electrons.

Ammonium ion, NH₄⁺

Figure 23 shows that NH_4^+ has a tetrahedral shape, with a bond angle of 109°.

Explanation of shape:

- electron pairs repel each other
- four bonding pairs of electrons repel each other equally
- molecule takes up a **tetrahedral** shape to minimise repulsions

Remember that ammonia (NH_3) is pyramidal (*three* bonding pairs of electrons and one lone pair of electrons), but an ammonium ion is tetrahedral (*four* bonding pairs of electrons).

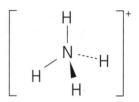

Figure 23 The shape of NH_4^+

More unusual examples

Exam tip

A double or a triple bond counts as one bonding pair of electrons when determining shape.

Carbon dioxide, CO₂

Figure 24 shows that the shape of CO_2 is linear with a bond angle of 180°.

Explanation of shape:

- electron pairs repel each other
- two sets of bonding pairs of electrons repel each other equally
- molecule takes a **linear** shape to minimise repulsions

$$O=C=O$$

Figure 24 The shape of CO_2

Bromine tetrafluoride ion, BrF₄⁻

Figure 25 shows that the shape of BrF_4^- is square planar, with a bond angle of 90°.

Explanation of shape:

- electron pairs repel each other
- four bonding pairs of electrons and two lone pairs of electrons
- lone pairs have greater repulsion than bonding pairs so the molecules takes up a **square planar** shape to minimise repulsions

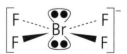

Figure 25 The shape of BrF_4^-

Bromine trifluoride, BrF₃

Figure 26 shows that BrF_3 is T shaped, with a bond angle of 86°.

Figure 26 The shape of BrF_3

Explanation of shape:

- electron pairs repel each other
- three bonding pairs of electrons and two lone pairs of electrons
- the basic shape is trigonal bipyramidal but lone pairs have greater repulsion than bonding pairs so the molecules takes a **T shape** to minimise repulsions

Xenon difluoride, XeF₂

Figure 27 shows that XeF_2 has a linear shape, with a bond angle of 180°.

Explanation of shape:

- electron pairs repel each other
- two bonding pairs of electrons and three lone pairs of electrons
- the basic shape is trigonal bipyramidal, but lone pairs have greater repulsion than bonding pairs so the molecules takes up a **linear** shape to minimise repulsions

Figure 27 the shape of XeF_2

> **Knowledge check 22**
>
> State the shape of an $AsCl_3$ molecule and the shape of a Cl_3^+ ion. To calculate the number of electrons for the Cl_3^+ ion, assume that one of the chlorine atoms is the central atom.

Summary

Table 12 shows the shapes of molecules and their bond angles.

Total number of electron pairs around central atom	Number of bonding pairs of electrons	Number of lone pairs of electrons	Shape	Bond angle	Examples
2	2	0	Linear	180°	$BeCl_2$, CO_2
3	3	0	Trigonal planar	120°	BF_3
4	4	0	Tetrahedral	109.5°	CH_4, NH_4^+
4	3	1	Pyramidal	107°	NH_3, H_3O^+
4	2	2	Bent	104.5°	H_2O, BrF_2^+
5	5	0	Trigonal bipyramidal	90° and 120°	PF_5
5	3	2	T shaped	86°	BrF_3
5	2	3	Linear	180°	XeF_2
6	6	0	Octahedral	90°	SF_6
6	4	2	Square planar	90°	BrF_4^-

Table 12 The shapes of molecules

Bond polarity

Electronegativity is the power of an atom to attract the pair of electrons in a covalent bond.

Trends in electronegativity

Electronegativity increases across a period. This is because the nuclear charge increases and the atomic radius decreases, whilst the shielding remains the same — the bonded electrons are closer to the attractive power of the nucleus.

Electronegativity decreases down a group. This is because the atomic radius increases and the shielding increases, hence the bonded electrons are further from the attractive power of the nucleus.

The most electronegative element is fluorine, which has electronegativity 4.0. The least electronegative element is caesium, which has electronegativity 0.7.

Polar bonds

In a covalent bond between two atoms which have the same, or very similar electronegativity, the bonding electrons are shared equally, i.e. electron distribution is symmetrical, for example, Cl–Cl, H–H, O=O, C–H.

In a covalent bond between two atoms that have a small difference in electronegativity, the electron pair is not shared equally but is displaced more towards the more electronegative atom, forming a **polar covalent bond**. This asymmetrical distribution of electrons is represented by the use of partial charges, δ+ (delta plus) and δ– (delta minus), above the atoms in the bond. The δ– is placed above the atom that has the higher electronegativity value and the δ+ is placed above the atom with the lower electronegativity value. The HCl bond is covalent but it has some ionic character (see Figure 28).

The difference in charge between the two atoms, caused by a shift in electron density in the bond, is called a dipole.

Large differences in electronegativity between atoms result in an ionic compound, for example NaCl, MgO, $CaBr_2$.

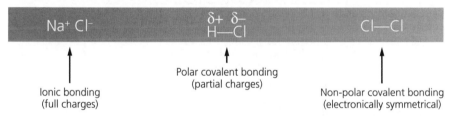

Figure 29 Electronegativity as a guide to bond polarity

Worked example

a Suggest why the bonding in carbon dioxide is covalent rather than ionic.

b Use partial charges to show the polarity of the C=O bond.

Answer

a There is a small difference in electronegativity between the values of carbon and oxygen.

b Oxygen is further across the period than carbon, and so it has a higher electronegativity, hence it will attract the electrons in the bond more, and become δ– (Figure 30).

Polarity of molecules

Simple covalent molecules can be polar or non-polar. The polarity depends on the presence of polar bonds, and also on their shape.

Knowledge check 23

Explain why the O–H bond is polar covalent.

$\overset{\delta+}{H}-\overset{\delta-}{Cl}$

Figure 28 Bond polarity of HCl

$\overset{\delta+}{C}=\overset{\delta-}{O}$

Figure 30 Bond polarity of C=O bond

If a molecule contains equally polar bonds arranged *symmetrically,* then the polarities of the bonds cancel each other out and the molecule is non-polar, for example, carbon dioxide, CO_2, is a linear molecule (Figure 31).

$$\overset{\delta-}{O} = \overset{\delta+}{C} = \overset{\delta-}{O}$$

Figure 31 Lack of polarity of CO_2

Generally, equally polar bonds arranged in a linear, trigonal planar, tetrahedral or octahedral arrangement create non-polar molecules.

In *asymmetrical* molecules such as water, the dipoles do not cancel and the molecule has a permanent dipole and is polar (Figure 32).

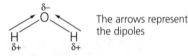
The arrows represent the dipoles

Figure 32 Polarity of water

Forces between molecules

There are three types of forces between molecules (intermolecular forces), which you need to be familiar with, as summarised in Table 13.

Type of intermolecular force	Where force occurs	Examples	Relative strength
Induced dipole–dipole forces (van der Waals forces, or London forces)	Between all atoms and molecules	Between non-polar molecules, for example I_2, Br_2, S_8, CCl_4; they are the *only* forces of attraction between these molecules	Weakest
Dipole–dipole forces	Between polar molecules	Between molecules of HCl or $CHCl_3$ or H_2S	
Hydrogen bonds	Between a $\delta+$ H atom (that is covalently bonded to O, N or F) in one molecule and an O, N or F atom in another molecule	Between molecules of NH_3, H_2O, HF	Strongest

Table 13

Induced dipole–dipole forces

These are also referred to as van der Waals, dispersion, or London forces.

Electrons are always moving in an atom, and this causes an unequal distribution of electrons, resulting in a temporary induced dipole. Induced dipole–dipole forces (Van der Waals forces) are attractions between these induced dipoles.

The more electrons that there are in a molecule, the greater the van der Waals forces of attraction between molecules — this explains the increase in boiling point of alkanes as the chain increases in length, and the increase in boiling point going down group 7.

Often, M_r is used as a measure of the number of electrons in a molecule. A larger molecule (greater M_r) has more electrons, and so demonstrates increased van der Waals forces between molecules.

Permanent dipole–dipole forces

Polar molecules contain atoms with different electronegativities and asymmetrical distribution of electrons — the molecules are said to have a **permanent dipole**. In addition to van der Waals forces they will have permanent dipole–dipole forces of attraction between the molecules. The permanent dipole results from the attraction between the $\delta+$ on one molecule and the $\delta-$ on another molecule.

Permanent dipole attractions exist between polar molecules like trichloromethane ($CHCl_3$), and propanone (CH_3COCH_3), as shown in Figure 33.

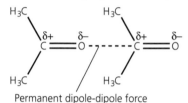

Permanent dipole-dipole force

Figure 33 Permanent dipole–dipole forces between propanone molecules

Exam tip

It is important to label the polarities ($\delta+$ and $\delta-$) of the polar bonds involved.

Hydrogen bonds

Hydrogen bonds are permanent dipole–dipole attractions between a $\delta+$ H atom (that is covalently bonded to O, N or F) in one molecule and a $\delta-$ O, N or F atom in another molecule. The hydrogen bond is formed because of the attraction between a lone pair of electrons on the $\delta-$ atom and the $\delta+$ hydrogen atom. Figure 34 shows the hydrogen bond between two water molecules.

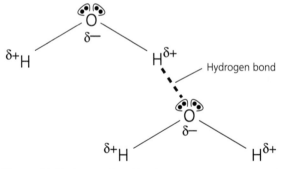

Figure 34 Hydrogen bonding between two water molecules

Exam tip

In diagrams between molecules to show hydrogen bonding, ensure that you show the polarity and all lone pairs. The lone pairs in this diagram are shown as (••). The hydrogen bond must be from the lone pair to the $\delta+$ H and is represented as a dashed line.

Using intermolecular forces to explain properties

Melting points and boiling points of molecular substances

When molecular substances are boiled or melted it is the intermolecular forces *between* the molecules that are broken, not the covalent bonds *within* them. When answering questions on melting point or boiling point you must always establish the type of intermolecular force(s) present.

Knowledge check 25

State the strongest type of intermolecular force in **a** water and **b** hydrogen sulfide (H_2S).

Explain why ethanol has a higher boiling point than propane, which has a similar M_r.

Answer

Propane only has relatively weak van der Waals forces between the molecules. There are strong hydrogen bonds between ethanol molecules (in addition to van der Waals forces) and it takes a large amount of energy to break these.

Explain why iodine has a higher boiling point than bromine.

Answer

I_2 has a higher M_r and so more electrons than Br_2, so there are increased van der Waals forces of attraction between the molecules. More energy is needed to break these increased intermolecular attractions and so iodine has a higher boiling point than bromine.

The anomalous boiling points of compounds

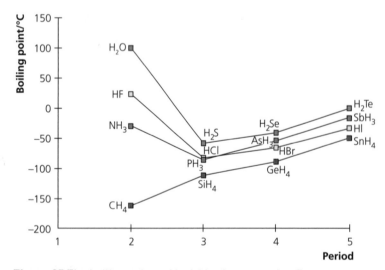

Figure 35 The boiling points of hydrides in groups 4 to 7

The boiling points of H_2O, HF and NH_3 are higher than expected. This is because of the strong hydrogen bonds between the molecules of these compounds. CH_4 is non-polar and so does not form hydrogen bonds between its molecules.

Note the increase in boiling point from H_2S to H_2Te in Figure 35 (and similar increases in the other groups). This increase in boiling point is due to increasing M_r and a higher number of electrons in the molecules, which increases the van der Waals forces of attraction between the molecules.

Explain why the boiling point of water is much higher than the boiling point of hydrogen sulfide.

Answer

There are strong hydrogen bonds between the water molecules (in addition to van der Waals forces) and it takes a large amount of energy to break these. Hydrogen sulfide has permanent dipole–dipole attractions between its molecules (in addition to van der Waals forces), which are weaker than hydrogen bonds.

Solubility in water

Liquids can be described as miscible (able to mix together in all proportions, forming one layer) or immiscible (unable to mix together, forming two distinct layers). Liquids often mix with each because of their ability to form the same intermolecular forces between the molecules. Solids are soluble in a solvent because they have similar bonds between their molecules. Many substances are insoluble in water because they cannot hydrogen bond with it.

Exam tip

'Like dissolves like' is a phrase that is often used to explain the miscibility of liquids and the solubility of a solid in a solvent. 'Like' in this case means that the bonding between the 'particles' in one substance is similar to the bonding between the 'particles' in the other substance. Bromine is non-polar and dissolves well in non-polar hexane.

Worked example

Explain why ethanol mixes with water.

Answer

Ethanol is soluble in water because it can form hydrogen bonds with water — hydrogen bonds can form between the lone pair of electrons on the $\delta-$ O and the $\delta+$ H (see Figure 36).

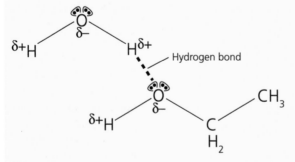

Figure 36 Hydrogen bond between ethanol and water

Explain why the boiling points of the group 6 hydrides increase from H_2S to H_2Se to H_2Te.

Exam tip

Remember to include all lone pairs and all partial charges in your diagram. The hydrogen bond must come from the lone pair.

Suggest why phosphine, PH_3, is insoluble in water, yet ammonia, NH_3, which has the same structure as phosphine, is soluble in water.

Viscosity

Viscosity is the opposite of fluidity (the more viscous a liquid the less well it flows).

> **Worked example**
>
> Explain why the liquid alkanes increase in viscosity as the carbon chain increases in length.
>
> **Answer**
>
> The number of electrons increases as the carbon chain length increases, so van der Waals forces are stronger between molecules, making the alkane less fluid in the liquid state.

The low density of ice

In liquid water the hydrogen bonds break and reform continuously as the molecules are moving. In ice:

- the hydrogen bonds are more ordered and hold the molecules in fixed positions in a three-dimensional tetrahedral lattice
- and so the water molecules are held further apart than they are in water

This more open arrangement means that ice has a lower density than water (fewer particles in the same space).

Bonding and physical properties

Crystals are solids with particles organised in a regular pattern. There are four types of crystal structure: ionic, metallic, molecular and macromolecular (giant covalent).

Ionic crystals

Sodium chloride is an example of an ionic crystal. The structure of sodium chloride is a giant ionic three-dimensional lattice of oppositely charged ions held by strong ionic bonds. It has a 6,6 crystal arrangement, which means that six Na^+ ions surround one Cl^- ion and six Cl^- ions surround one Na^+ ion. It has a cube shape (Figure 37).

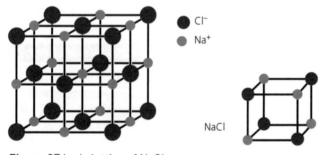

Figure 37 Ionic lattice of NaCl

The physical properties of ionic crystals such as sodium chloride are explained in Table 14.

> **Exam tip**
>
> You are expected to be able to draw diagrams to represent an ionic structure — a simple cube such as the one on the right in Figure 37 is sufficient. Remember to have oppositely charged ions beside each other. You can simply write Na^+ and Cl^- inside each circle, instead of colour coding.

Physical property	Explanation of physical property in terms of structure and bonding
Crystalline	Regular arrangement of positive and negative ions causes the crystalline structure
High melting point and boiling point	Large amount of energy is required to break the bonds, which are strong electrostatic attractions between ions of opposite charge
Non-conductor of electricity when solid	Ions are not free to move and cannot carry charge
Good conductor of electricity when molten or when aqueous (dissolved in water)	Ions are free to move and can carry charge

Table 14

Metallic crystals

A metallic crystal consists of layers of positive ions held by a sea of delocalised electrons. Magnesium is an example of a metallic crystal. An explanation of the physical properties of metallic crystals is shown in Table 15.

Physical property	Explanation of physical property in terms of structure and bonding
Hardness	Strong attraction between positive ions and negative electrons, and a regular structure
High melting point	Large amount of energy is required to break the bonds, which are strong attractions between positive ions and negative electrons
Good electrical conductivity	Delocalised electrons can move and carry charge through the metal
Malleable (can be hammered into shape) and ductile (can be drawn into a wire)	Layers of positive ions can slide over each other without disrupting the bonding

Table 15

Molecular crystals

Iodine, I_2, is an example of a molecular crystal. Its molecules are held together in a regular arrangement (Figure 38) usually by weak intermolecular forces. In iodine the intermolecular forces between the molecules are induced dipole–dipole attractions.

Table 16 explains the physical properties of molecular crystals such as iodine.

Physical property	Explanation of physical property in terms of structure and bonding
Low melting point	It does not take much energy to break the weak induced dipole–dipole forces between the molecules
Non-conductor of electricity	The electrons are all used in bonding and are not free to move and carry the charge
Insoluble in water	They are non-polar and do not interact with water. Any which do appear to dissolve in water, for example, Cl_2, are usually reacting with the water rather than dissolving
Soft	Surface atoms are held weakly by intermolecular bonds and so can be removed easily by force

Table 16

Ice is another example of a molecular crystal. In ice the intermolecular forces are hydrogen bonds (see page 50).

Exam tip

In an exam you could be asked to draw a diagram to show how the particles are arranged in any metal — draw a section of the diagram shown in Figure 14. Remember that the total number of delocalised electrons should equal the total charge of the cations.

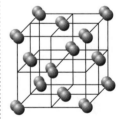

Figure 38 Iodine crystal

Macromolecular crystals (giant covalent)

Diamond and graphite are allotropes of carbon — they are different structures of the same element.

- In diamond each carbon atom is covalently bonded to four others in a tetrahedral arrangement (Figure 39). The structure is a rigid three-dimensional lattice.
- In graphite there are hexagonal layers of carbon atoms, each connected by covalent bonds (Figure 40). Between the layers there are weak forces. Carbon atoms have four unpaired electrons and can form four covalent bonds. Each carbon is covalently bonded to three others, so each carbon has one electron available which is delocalised and free to move between the layers.

Exam tip

When a molecular crystal is melted or boiled it is the intermolecular forces between molecules which break, *not* the covalent bonds within the molecule.

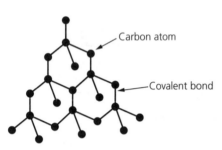

Carbon atom

Covalent bond

Figure 39 Structure of diamond

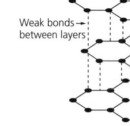

Covalent bond

Carbon atom

Weak bonds → between layers

Figure 40 Structure of graphite

Physical property	Explanation
Graphite and diamond have very high melting points and boiling points	A large amount of energy is needed to break the many strong covalent bonds between atoms
Diamond is very hard	There are many strong covalent bonds and a rigid three-dimensional structure holding surface atoms in place. Diamond is used in drills and cutting tools
Graphite is very soft	There are weak forces between the layers, which means that the layers can slide over each other. This accounts for the soft, flakiness of graphite and its use in pencil lead and as a lubricant
Diamond does not conduct electricity	All of the carbon's four valence electrons are localised in bonds. None are free to move throughout the molecule
Graphite conducts electricity	Only three of carbon's valence electrons are used in covalent bonding — the fourth electron becomes delocalised. The delocalised electrons can move and carry charge

Table 17

Energy changes associated with changes of state

To explain energy changes associated with change of state you need to understand the arrangement of particles in the three states of matter as shown in Table 18.

Melting

Melting is the change of state from solid to liquid.

An increase in energy gives the particles more energy and they vibrate faster. At the melting point the temperature remains constant because the energy is used to break the forces of attraction between the particles and cause them to move apart — this is the enthalpy change of fusion.

Knowledge check 28

Explain why graphite has a high melting point and phosphorus has a low melting point.

	Solid	Liquid	Gas
Order	High degree of order. Particles are close together	Intermediate. Particles are further apart than in a solid	Disordered. Particles are far apart
Movement of particles	Particles vibrate about a fixed position	Some movement	Rapid and random
Forces of attraction between particles	Strong	Weaker than in a solid	Negligible

Table 18

Boiling

Boiling is the change of state from liquid to gas.

An increase in energy gives the particles more energy and they move faster and move further apart. At the boiling point the temperature remains constant because the energy is used to break the remaining forces of attraction between the particles and cause them to move apart — this is enthalpy of vaporisation.

Freezing

Freezing is the change of state from liquid to solid.

When bonds form, energy is released. The stronger the bonds formed on freezing, the more energy is released.

Condensing

Condensing is the change of state from gas to liquid.

Energy is released when a substance condenses. The stronger the bonds formed on condensing, the more energy is released.

Subliming

Subliming is the change of state from solid straight to gas on heating (without passing through the liquid stage) or from gas straight to solid on cooling.

Substances that sublime are solid iodine and solid carbon dioxide, which is known as dry ice. Iodine sublimes from a grey-black solid to a purple gas.

Summary

- Most compounds containing a metal exhibit ionic bonding (electrostatic attractions between oppositely charged ions). Exceptions are $BeCl_2$ and $AlCl_3$.
- Compounds containing only non-metals exhibit covalent bonding — a single covalent bond is a shared pair of electrons.
- Metallic bonding involves attraction between delocalised electrons and positive ions arranged in a lattice.
- Electronegativity is the power of an atom in a covalent bond to attract the bonding electrons. If the electron distribution in a bond is asymmetrical a polar covalent bond forms with $\delta+$ and $\delta-$ partial charges.

- Simple covalent molecules can be described as polar or non-polar based on the difference in electronegativity of the atoms involved and also the symmetry of the molecule.
- The three types of intermolecular forces of attraction between simple molecules are van der Waals (induced dipole-dipole) forces, permanent dipole-dipole attractions and hydrogen bonds.
- Non-polar molecules and simple atoms (like the noble gases) have only van der Waals forces of attraction between their particles. Van der Waals forces of attraction are caused by induced dipoles, which are the result of the random movement of electrons in the molecule. The greater the number of electrons in the molecule, the greater the van der Waals forces of attraction.
- Polar molecules have permanent dipole–dipole attractions between the molecules, as well as van der Waals forces of attraction.

- Hydrogen bonds are attractions between a δ+ H atom (that is covalently bonded to O, N or F) in 1 molecule and a δ– O, N or F atom in another molecule. The hydrogen bond is formed because of the attraction between a lone pair of electrons on the δ– atom and the δ+ hydrogen atom. Hydrogen bonds are the strongest of the intermolecular forces of attraction.
- Physical properties of simple covalent elements or compounds can be explained in terms of the intermolecular forces between their molecules.
- There are four types of crystal structure: ionic (for example, sodium chloride); metallic (for example, magnesium); molecular (for example, iodine, ice); and macromolecular (for example, graphite, diamond).
- The physical properties of substances can be explained in terms of their structure and bonding.

■ Energetics

Enthalpy is the total thermodynamic energy in a system. The system can be the reactants or the products in a chemical reaction.

ΔH is the **enthalpy change,** and the units of H and ΔH are kJ.

Chemical reactions are described as either exothermic or endothermic.

- For an exothermic reaction, heat is given out to the surroundings and ΔH is negative (the enthalpy of the products is less than the enthalpy of the reactants).
- For an endothermic reaction heat is taken in from the surroundings and ΔH is positive (the enthalpy of the products is more than the enthalpy of the reactants).

The **enthalpy change** (ΔH) is the heat energy change measured under conditions of constant pressure.

Enthalpy change

Enthalpy level diagrams show the relative enthalpy of the reactants and products together with the change in enthalpy. The diagrams for an endothermic reaction and for an exothermic reaction look different. As energy is taken in during an endothermic reaction the products are at a higher enthalpy value so the change in enthalpy (ΔH) is positive (+ve). As energy is released during an exothermic reaction the enthalpy decreases from reactants to products so the change in enthalpy (ΔH) is negative (–ve). These enthalpy changes are shown in Figure 41.

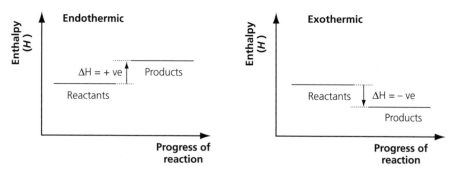

Figure 41 Enthalpy profiles of endothermic and exothermic reactions

Changes of state

Endothermic changes of state are melting, boiling, evaporation and sublimation, as energy is taken in to cause the change in state.

Exothermic changes of state are condensation and freezing, as energy is released during these changes of state.

Standard enthalpy changes

Standard enthalpy changes are the enthalpy changes for specific reactions that occur under standard conditions. Standard conditions are 100 kPa and a stated temperature, usually 298 K (25°C).

Standard enthalpy changes are measured in kJ mol^{-1} (kilojoules per mole) and they are measured relative to 1 mole of a particular substance in the reaction. When an enthalpy change is under standard conditions, the $\ominus$ symbol is used as a superscript after ΔH.

Standard enthalpy changes at 298 K are sometimes written as $\Delta H_{298}^{\ominus}$.

> **Exam tip**
>
> The 'per mol' concept is important for calculations as you can work out the energy change in a reaction based on standard enthalpy changes and the number of moles of the particular substance. The standard enthalpy change of combustion of ethanol is –1367 kJ mol^{-1}, which means that 1367 kJ of energy is released (negative sign means exothermic) when 1 mol of ethanol is burned completely in oxygen. If 0.1 mol of ethanol is burned completely in oxygen, 136.7 kJ of energy would be released.

When writing equations for reactions representing standard enthalpy changes, you must include state symbols for all reactants and products. All substances in the reaction are in their standard states, at 298 K and 100 kPa pressure.

> **Exam tip**
>
> Make sure that you can label an enthalpy level diagram correctly, including the axes labels, reactants and products, and show the correct increase or decrease in enthalpy. If the diagram is for a specific reaction label the reactants and products lines with the appropriate reactants and products. The 'progress of reaction' axis may also be labelled 'reaction coordinate'.

> **Knowledge check 29**
>
> What is meant by ΔH?

Standard enthalpy change of combustion

Symbol: $\Delta_c H^\ominus$ Units: $kJ\,mol^{-1}$

The equation representing the **standard enthalpy change of combustion** of methane is:

$$CH_4(g) + 2O_2(g) \rightarrow CO_2(g) + 2H_2O(l)$$

1 mole of methane is burned completely (forming carbon dioxide and water) and all substances are in their standard states at 25°C and 100 kPa. $\Delta_c H^\ominus$ (methane) = $-890\,kJ\,mol^{-1}$. This means that when 1 mole of methane is burned completely in oxygen 890 kJ of energy are released.

The equation representing the standard enthalpy change of combustion of ethane is:

$$C_2H_6(g) + 3\tfrac{1}{2}O_2(g) \rightarrow 2CO_2(g) + 3H_2O(l)$$

1 mole of ethane is burned completely and all substances are in their standard states:

$\Delta_c H^\ominus$ (ethane) = $-1560\,kJ\,mol^{-1}$

Exam tip

Don't be afraid to use ½ and other fractions to balance an equation representing any standard enthalpy change, as it must be written for 1 mole of the substance to which it relates. In the combustion examples given it has to be 1 mole of methane and 1 mole of ethane and the other substance are balanced accordingly.

Standard enthalpy change of neutralisation

Symbol: $\Delta_n H^\ominus$ **Units:** $kJ\,mol^{-1}$

The equation representing the **standard enthalpy change of neutralisation** is:

$$H^+(aq) + OH^-(aq) \rightarrow H_2O(l)$$

1 mole of water is formed in a neutralisation reaction and all substances are in their standard states.

The standard enthalpy change of neutralisation for a strong acid reacting with a strong alkali is $-57.6\,kJ\,mol^{-1}$. For a weak acid, such as ethanoic acid, reacting with a strong alkali, such as sodium hydroxide solution, the value is $-56.1\,kJ\,mol^{-1}$. This value is lower as the weak acid is not completely dissociated and some energy is required to complete this dissociation, so less energy is released as heat.

Standard enthalpy change of formation

Symbol: $\Delta_f H^\ominus$ **Units:** $kJ\,mol^{-1}$

The equation representing the **standard enthalpy change of formation** of ethanol is:

$$2C(s) + 3H_2(g) + \tfrac{1}{2}O_2(g) \rightarrow C_2H_5OH(l)$$

1 mole of ethanol is formed from its elements under standard conditions, and all substances are in standard states:

$\Delta_f H^\ominus$ (ethanol) = $-277.7\,kJ\,mol^{-1}$

The **standard enthalpy change of combustion** is the enthalpy change when 1 mole of a substance is burned completely in oxygen and all reactants and products are in their standard states under standard conditions.

The **standard enthalpy change of neutralisation** is the standard enthalpy change when 1 mole of water is produced in a neutralisation reaction under standard conditions.

The **standard enthalpy change of formation** is the enthalpy change when 1 mole of a substance is formed from its constituent elements, and all reactants and products are in their standard states under standard conditions.

The equation representing the standard enthalpy change of formation of sodium nitrate is:

$$Na(s) + \frac{1}{2}N_2(g) + 1\frac{1}{2}O_2(g) \rightarrow NaNO_3(s)$$

1 mole of sodium nitrate is formed from its elements under standard conditions, and all substances are in standard states:

$$\Delta_f H^\ominus (NaNO_3) = -466.7 \, kJ \, mol^{-1}$$

Standard enthalpy change of reaction

Symbol: $\Delta_r H^\ominus$ Units: $kJ \, mol^{-1}$

The equation representing the **standard enthalpy of reaction** for the hydrogenation of ethene is:

$$C_2H_4(g) + H_2(g) \rightarrow C_2H_6(g)$$

$$\Delta_r H^\ominus = -138 \, kJ \, mol^{-1}$$

The units of standard enthalpy changes of reaction may be given as kJ or $kJ \, mol^{-1}$. $kJ \, mol^{-1}$ are almost always used.

Calorimetry

Temperature change may be converted to energy change using the expression:

$$q = mc\Delta T$$

where q = change in energy in joules; m = mass in grams of the substance to which the temperature change occurs — usually water (for combustion) or solution (for neutralisation); c = specific heat capacity (the energy required to raise the temperature of $1.00 \, g$ of a substance by $1°C$); and ΔT = temperature change in $°C$ or kelvin (K).

Calculating enthalpy change of neutralisation

The apparatus used for determining the enthalpy change of neutralisation is shown in Figure 42.

1 Add an exact volume of a known concentration of alkali into a polystyrene cup and measure the initial temperature.

2 Add an exact volume of a known concentration of acid to the polystyrene cup.

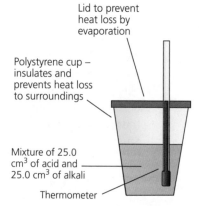

Figure 42 Apparatus used to measure enthalpy changes in solution

Lid to prevent heat loss by evaporation

Polystyrene cup – insulates and prevents heat loss to surroundings

Mixture of 25.0 cm³ of acid and 25.0 cm³ of alkali

Thermometer

The **standard enthalpy change of reaction** is the enthalpy change when substances react under standard conditions with the number of moles given by the equation for the reaction.

Knowledge check 30

What are the standard conditions used in enthalpy changes?

Exam tip

For water the specific heat capacity = $4.18 \, J \, K^{-1} \, g^{-1}$. For many solutions in neutralisation reactions the value is assumed to be the same.

3   Use a thermometer to measure the highest temperature achieved.
    Calculate the temperature change (ΔT).

4   Calculate the energy change using $q = mc\Delta T$.

5   Calculate the energy change for 1 mole of water by dividing by the
    number of moles of water produced in the neutralisation reaction.

Calculating enthalpy change of combustion

The apparatus used for determining the enthalpy change of combustion is shown in
Figure 43.

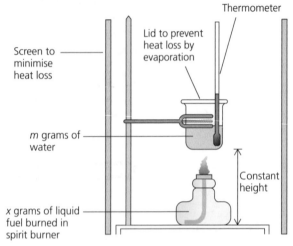

Figure 43 Apparatus used to measure enthalpy changes of combustion

1   Measure the mass of fuel that is burned. This is most often achieved by
    measuring the mass of the spirit burner before and after combustion.

2   Measure the initial temperature of the water.

3   Allow the fuel to burn in a spirit burner and to heat a known volume of water.

4   Use a thermometer to measure the highest temperature achieved.
    Calculate the temperature change (ΔT).

5   Calculate the energy change using $q = mc\Delta T$.

6   Calculate the energy change for 1 mole of fuel by dividing by the
    number of moles of fuel burned.

Typical calculations

Worked example 1

500 cm^3 of water were heated by burning 3.16 g of ethanol in a spirit burner. The
initial temperature of the water was 20.5°C and the highest temperature recorded
was 61.1°C. The specific heat capacity of water is 4.18 J K^{-1} g^{-1}. Calculate the
standard enthalpy change of combustion of ethanol.

Exam tip

The density of water is
1.00 g cm^{-3} so 100 cm^3
of water = 100 g.

Exam tip

When enthalpy values
are calculated from
experimental data,
the values may be
less than expected, as
some heat is lost to the
surroundings and also
by evaporation.

Exam tip

For any calculation
where you know a value
for a certain amount
in moles, always
divide the value by the
number of moles to
calculate it per mole
of a substance even if
the amount in moles is
smaller than 1.

Answer

$\Delta T = 61.1 - 20.5 = 40.6°C \ (= 40.6\,K)$

$m = 500\,g$ (mass of water as volume of water = $500\,cm^3$)

$c = 4.18\,J\,K^{-1}\,g^{-1}$

$q = mc\Delta T = 500 \times 4.18 \times 40.6 = 84\,854\,J$

mass of ethanol = $3.16\,g$

moles of ethanol $= \dfrac{3.16}{46.0} = 0.0687$

per mole of ethanol $= \dfrac{84\,854}{0.0687} = 1\,235\,138.28\,J$

standard enthalpy change of combustion $= \dfrac{1\,235\,138.28}{1000} = -1235.138\,kJ\,mol^{-1}$

Worked example 2

0.0540 moles of water were formed in a neutralisation reaction. The solution volume was $50.0\,cm^3$. The temperature increase recorded was $13.1°C$. Assuming that the specific heat capacity of the solution is $4.18\,J\,K^{-1}\,g^{-1}$ and that the density of the solution is $1.00\,g\,cm^{-3}$, calculate the standard enthalpy change of neutralisation. Give your answer to 3 significant figures.

Answer

$\Delta T = 13.1°C$

$m = 50.0\,g$ (using density)

$c = 4.18\,J\,K^{-1}\,g^{-1}$

$q = mc\Delta T = 50.0 \times 4.18 \times 13.1 = 2737.9\,J$

moles of water formed = 0.0540

ΔE (change in energy) per mol of water formed $= \dfrac{2737.9}{0.0540} = 50\,701.852\,J$

standard enthalpy change of neutralisation $= \dfrac{50701.852}{1000} = -50.7\,kJ\,mol^{-1}$

Required practical 2

Measurement of an enthalpy change
This practical may ask you to determine the enthalpy change for a solid dissolving in water or for a neutralisation or displacement reaction or the combustion of an alcohol. To carry out this practical you must be familiar with the content on pages 54 to 59.

Exam tip

The units of standard enthalpy changes are $kJ\,mol^{-1}$. The negative sign is put in at the end before the value as it is exothermic. This is a common omission by students and can cost you a mark. If you calculate an enthalpy change from experimental data, always think about whether it is exothermic or endothermic and place the appropriate sign in front of the value.

Knowledge check 31

In $q = mc\Delta T$ what do m, c and ΔT represent? State their units.

Content Guidance

Applications of Hess's law

Hess's law restates the **principle of conservation of energy**.

Hess's law can be used to calculate enthalpy changes for chemical reactions from the enthalpy changes of other reactions. This is useful as some reactions cannot be carried out, but a theoretical enthalpy change can be determined for them.

Standard enthalpy of formation values can be used to calculate an enthalpy of reaction or an enthalpy of combustion:

$$\Delta_r H^{\ominus} \text{ or } \Delta_c H^{\ominus} = \Sigma\Delta_f H^{\ominus}(\text{products}) - \Sigma\Delta_f H^{\ominus}(\text{reactants})$$

where Σ is the sum of the enthalpies of formation, taking into account the number of moles of a reactant or product.

Standard enthalpies of combustion values can be used to calculate an enthalpy of formation:

$$\Delta_f H^{\ominus} = \Sigma\Delta_c H^{\ominus}(\text{reactants}) - \Sigma\Delta_c H^{\ominus}(\text{products})$$

where Σ is the sum of the enthalpies of combustion, taking into account the number of moles of a reactant or product.

Worked example 1

Calculate the standard enthalpy of formation of butanone, $C_4H_8O(l)$, given the following standard enthalpies of combustion:

$C(s)$	$\Delta_c H^{\ominus} = -394\,kJ\,mol^{-1}$
$H_2(g)$	$\Delta_c H^{\ominus} = -286\,kJ\,mol^{-1}$
$C_3H_6O(l)$	$\Delta_c H^{\ominus} = -2444\,kJ\,mol^{-1}$

Answer

In this example the values given are standard enthalpies of combustion and the value to be calculated is an enthalpy of formation.

The equation that represents the enthalpy of formation of butanone is:

$$4C(s) + 4H_2(g) + \tfrac{1}{2}O_2(g) \rightarrow C_4H_8O(l)$$

The expression shown in Figure 44 should be used.

$\Delta_f H = \Sigma\Delta_c H(\text{reactants}) - \Sigma\Delta_c H(\text{products})$
$\Delta_f H = +4(-394) + 4(-286) - (-2444) = -276\ kJ\ mol^{-1}$

4 here because 4 moles of C(s) react	4 here because 4 moles of H$_2$(g) react	The –(–2444) is +2444 because this enthalpy change is being reversed

Figure 44

The **principle of conservation of energy** states that energy cannot be created or destroyed, only changed from one form into another.

Hess's law states that the enthalpy change for a chemical reaction is independent of the route (or the number of steps) taken, provided that the initial and final conditions remain the same.

Essentially, the enthalpy change for the overall reaction is the same as the enthalpy change for the reactants burning to form CO_2 and H_2O and reversing the burning of the products. Figure 45 shows this process in a Hess's law diagram. The combustion of 4 moles of $C(s)$ to form 4 moles of $CO_2(g)$ is $4(-394)$; the combustion of 4 moles of $H_2(g)$ to form 4 moles of $H_2O(l)$ is $4(-286)$. Adding in the reverse of the combustion of 1 mole of butanone makes the overall change:

$$4C(s) + 4H_2(g) + \frac{1}{2}O_2(g) \rightarrow C_4H_8O(l)$$

Figure 45 Hess's law diagram for the formation of butanone

$$\Delta_fH^\ominus = (4 \times -394) + (4 \times -286) - (-2444) = -276\,\text{kJ}\,\text{mol}^{-1}$$

Knowledge check 32

The enthalpies of combustion of $C(s)$, $H_2(g)$ and $C_4H_{10}(g)$ are -394, -286 and $-2878\,\text{kJ}\,\text{mol}^{-1}$. Calculate the standard enthalpy of formation of butane.

Worked example 2

Calculate the standard enthalpy change of combustion of propane from the following standard enthalpies of formation:

$CO_2(g)$	$\Delta_fH^\ominus = -394\,\text{kJ}\,\text{mol}^{-1}$
$H_2O(l)$	$\Delta_fH^\ominus = -286\,\text{kJ}\,\text{mol}^{-1}$
$C_3H_8(g)$	$\Delta_fH^\ominus = -105\,\text{kJ}\,\text{mol}^{-1}$
$O_2(g)$	$\Delta_fH^\ominus = 0\,\text{kJ}\,\text{mol}^{-1}$

Main equation:

$$C_3H_8(g) + 5O_2(g) \rightarrow 3CO_2(g) + 4H_2O(l)$$

$$\Delta_cH^\ominus = \Sigma\Delta_fH^\ominus(\text{products}) - \Sigma\Delta_fH^\ominus(\text{reactants})$$

$$\Delta_cH^\ominus = 3(-394) + 4(-286) - (-105) = -2221\,\text{kJ}\,\text{mol}^{-1}$$

Essentially, the enthalpy change for the overall reaction is the reverse of the enthalpy change for the reactants being converted into their elements and the elements being used to form the products. The Hess's Law diagram in Figure 46 shows this process. 1 mole of propane is theoretically converted to its elements (reverse of formation) and the 3 moles of $C(s)$ are converted to 3 moles of $CO_2(g)$ ($3 \times$ enthalpy of formation of CO_2) and 4 moles of $H_2(g)$ are converted to 4 moles of $H_2O(l)$ ($4 \times$ enthalpy of formation of H_2O).

Exam tip

The reactant $O_2(g)$ is already an element in its standard states so its enthalpy of formation is zero.

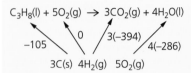

Figure 46 Hess's law diagram for the combustion of propane

Bond enthalpies

The mean bond enthalpy is the energy required to break 1 mole of a covalent bond in the gaseous state, measured in $kJ\,mol^{-1}$. The 'per mole' is per mole of the covalent bond averaged across many compounds containing the bond. For example, the mean bond enthalpy of O–H is $463\,kJ\,mol^{-1}$.

Remember bond breaking is endothermic and bond making is exothermic.

Breaking 1 mole of O–H $+463\,kJ\,mol^{-1}$ (+ve as endothermic)

Making 1 mole of O–H $-463\,kJ\,mol^{-1}$ (−ve as exothermic)

The term mean bond enthalpy is used as the enthalpy of 1 mole of the covalent bond is averaged across different molecules containing the bond.

Typical calculation

Worked example

Determine the standard enthalpy of combustion of ethene (C_2H_4) using the mean bond enthalpy values shown in Table 19.

Bond	Mean bond enthalpy ($kJ\,mol^{-1}$)
C–H	413
C=C	611
O–H	464
O=O	497
C=O	803

Table 19

Answer

The equation for the reaction should be written as a balanced symbol equation (remember that it should be per mole of the fuel burned (ethene) to be a standard enthalpy change of combustion).

$$C_2H_4(g) + 3O_2(g) \rightarrow 2CO_2(g) + 2H_2O(g) \quad \text{(per mole of } C_2H_4\text{)}$$

Then draw a structural equation showing all the covalent bonds, as shown in Figure 47:

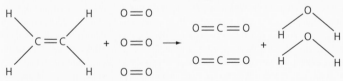

Figure 47 Covalent bonds in the combustion of ethene

Knowledge check 33

Calculate the standard enthalpy of combustion of ethanol given the following enthalpies of formation:
$C_2H_5OH(l)$ $-279\,kJ\,mol^{-1}$;
$CO_2(g)$ $-394\,kJ\,mol^{-1}$;
$H_2O(l)$ $-286\,kJ\,mol^{-1}$.

Exam tip

Calculations involving average bond enthalpy values often differ from the actual values due to the fact that average bond enthalpy values are not specific to the molecules in the reaction or the molecules in the reaction may not be in the gaseous state.

Calculate the energy required for moles of bond broken in the reactants:

1 C=C		= 611 kJ
4 C–H	4(413)	= 1652 kJ
3 O=O	3(497)	= 1491 kJ

total energy required for bonds broken = 3754 kJ

Calculate the energy released for moles of bonds formed in the products:

4 C=O	4(803)	= 3212 kJ
4 O–H	4(464)	= 1856 kJ

total energy released for bonds made = 5068 kJ

The overall energy change is the energy required to break bonds in reactants – energy released when bonds form in products. The energy required to break the bonds is positive (+ve) as it is endothermic and the energy released when bonds form is negative (–ve) as it is exothermic.

overall energy change for reaction = +3754 – 5068 = –1314 kJ mol^{-1}

The standard enthalpy of combustion of ethene is quoted as –1411 kJ mol^{-1}. The difference is caused by the mean bond enthalpies being averaged across molecules containing the bond; they are not specific to the molecules in this reaction. Also, the substances for mean bond enthalpies are all gases, whereas in the standard enthalpy of combustion of ethene, water would be in its standard state as a liquid.

Knowledge check 34

Which of the following bonds would have the lowest bond enthalpy value: H–Cl, H–Br, H–I?

Summary

- ΔH is the change in enthalpy in a system (reaction). Where ΔH is positive the reaction is endothermic and where ΔH is negative the reaction is exothermic.
- An enthalpy level diagram shows enthalpy against progress of reaction (reaction coordinate).
- Standard enthalpy changes of combustion, neutralisation, formation and reaction are all under standard conditions of 25°C (298 K) and 100 kPa pressure.
- Temperature changes from experiments can be used to calculated enthalpy changes using $q = mc\Delta T$.
- Hess's law allows calculation of an enthalpy change using other standard enthalpy changes.
- Bond breaking is endothermic and bond making is exothermic.
- Mean bond enthalpy values can also be used to calculate an enthalpy change using the energy of the covalent bonds broken and covalent bonds made in the reaction.

■ Kinetics

Chemical kinetics is the study of rates of reaction.

The rate of reaction is a measure of the change in the concentration (amount) of a reactant or product with respect to time.

Measuring rate of reaction

The rate of a reaction may be determined by measuring change in the volume of a gas, change in mass, or mass against time. Many of the reactions used to study kinetics at this level produce a gas. Graphs of gas volume or mass against time can be drawn. Changes in the other quantities, such as concentration or temperature, have an effect on the shape of the graphs. The following two examples illustrate the point.

The graph in Figure 48 shows the change in mass for the reaction between calcium carbonate and hydrochloric acid carried out at two different temperatures, 25°C and 50°C. The equation for the reaction is:

$$CaCO_3 + 2HCl \rightarrow CaCl_2 + CO_2 + H_2O$$

The same mass of calcium carbonate, and volume and concentration of hydrochloric acid, is used for each reaction.

The experiments are carried out in a conical flask on an electronic balance.

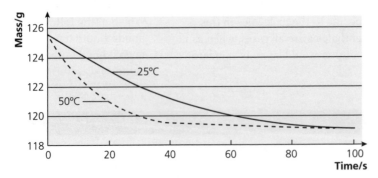

Figure 48 Graph of mass against time at different temperatures

The reaction at 50°C is faster than the reaction at 25°C, so the mass decreases much more rapidly at 50°C. As the mass of calcium carbonate and the volume and concentration are the same in each experiment, the initial mass and final mass are the same. The actual curves from initial mass to final mass are different for each temperature.

The graph in Figure 49 shows gas volume against time for the reaction between calcium carbonate and hydrochloric acid, for two different concentrations of hydrochloric acid, where the acid is in excess. The same mass of calcium carbonate is used for each reaction. The reactions are carried out using a gas syringe to measure the gas volume, and the graphs show the gas volume against time for the two different concentrations of hydrochloric acid, at a constant temperature. The reaction at the higher concentration of hydrochloric acid produces gas more rapidly, but both reactions start and end at the same gas volume.

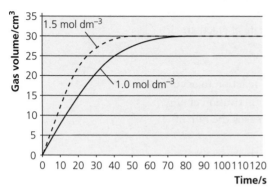

Figure 49 Graphs of gas volume against time at different concentrations

Required practical 3

Investigate how the rate of reaction changes with temperature
This practical may ask you to investigate the effect of temperature on the rate of reaction of sodium thiosulfate and hydrochloric acid, by an initial rate method. To carry out this practical you must be familiar with the content on pages 64 to 65.

Collision theory

Reaction rate depends on the number of successful collisions in a given period of time. The number of collisions can be affected by changes in temperature, pressure, concentration and the presence of a catalyst. A successful collision is a collision in which the particles have at least the minimum energy required for a reaction to occur. The minimum energy required for a reaction to occur is called the **activation energy**.

Effect of temperature on reaction rate

Increasing temperature increases the energy of reacting particles, which leads to an increase in the frequency of collisions and the frequency of successful collisions, which increases the rate of reaction.

Effect of pressure and concentration on reaction rate

Pressure

Increasing the pressure pushes the reacting particles closer together, which increases the frequency of collisions. This leads to an increase in the frequency of successful collisions, which increases the rate of reaction.

Concentration

Increasing the concentration of the reactant(s) increases the number of reacting particles, which results in an increased frequency of collisions. This leads to an increased frequency of successful collisions in a given period of time, which increases the rate of reaction.

Catalysts

Activation energy is the minimum energy that reactants need for a reaction to occur.

- A catalyst provides an alternative reaction route that has a lower activation energy, which increases the number of successful collisions and so increases the rate of reaction.
- A homogeneous catalyst is a catalyst that is in the same state as the reactants and products, and provides a reaction route of lower activation energy by forming an intermediate.
- A heterogeneous catalyst is a catalyst that is in a different state to the reactants and products, and works via chemisorption to provide a reaction route of lower activation energy.
- Chemisorption is the process by which reactant molecules are adsorbed onto the surface of the catalyst, bonds are weakened and product molecules are desorbed from the catalyst.

The size of any solid reacting particles also has an effect on the rate of the reaction as smaller solid particles cause the reaction to occur at an increased rate. Powdered magnesium will react faster with hydrochloric acid than magnesium ribbon.

Enthalpy level diagrams

In the Energetics section we considered an enthalpy profile diagram for an exothermic and an endothermic reaction (Figure 41). We now have to include the reaction route into the diagram and be able to recognise and/or label the activation energy in the diagram. Figure 50 is an enthalpy level diagram for an exothermic reaction.

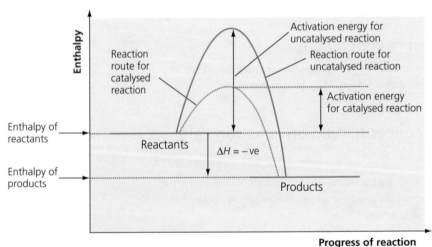

Figure 50 Enthalpy level diagram showing activation energy. The activation energy for the catalysed reaction is lower than the activation energy for the uncatalysed reaction

---

Knowledge check 35

Explain how a catalyst increases the rate of reaction.

Exam tip

Often you will be asked to sketch an enthalpy profile diagram and this should not be confused with a Maxwell–Boltzmann distribution.

A catalyst for an endothermic reaction also provides an alternative reaction route of lower activation energy.

Maxwell–Boltzmann distribution

Also called distribution of molecular energies, a Maxwell–Boltzmann distribution is a plot of number of gaseous molecules against the energy they have (Figure 51). It should appear as a roughly normal distribution, which is asymptotic to the horizontal axis (gets closer and closer but never touches the axis) at higher energy.

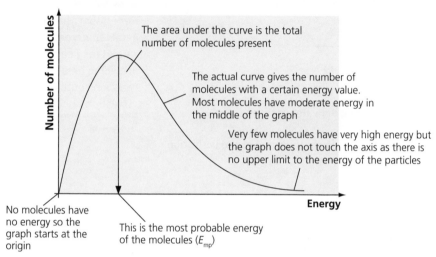

Figure 51 A Maxwell–Boltzmann distribution

The most probable energy is often labelled E_{mp}. The value of E_{mp} increases when temperature increases but does not change when concentration changes or a catalyst is added.

Activation energy on a Maxwell–Boltzmann distribution

The activation energy (E_a) is the minimum energy that gaseous reactant molecules must possess to undergo a reaction. It is the top of the hump on an enthalpy profile diagram. If the Maxwell–Boltzmann distribution represents the energy of the reactant molecules then there will be an energy value on the x-axis that is the activation energy.

Knowledge check 36

What labels are placed on the vertical and horizontal axes of a Maxwell–Boltzmann distribution?

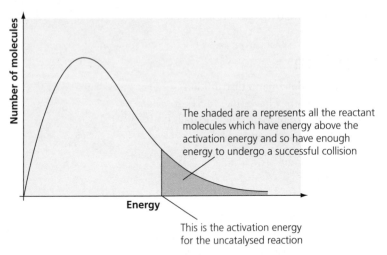

Figure 52 Maxwell–Boltzmann distribution showing activation energy

Exam tip

Remember that a catalyst increases the rate of the reaction by providing an alternative reaction route of lower activation energy.

Adding a catalyst to the reaction in Figure 52 would lower the value of the activation energy, as shown in Figure 53.

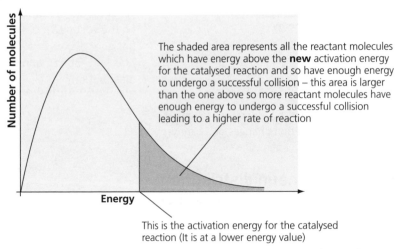

Figure 53 Maxwell-Boltzmann distribution in the presence of a catalyst

Maxwell–Boltzmann distribution at different temperatures

When the temperature of the reactant molecules is increased, this will increase the energy of the gaseous reactant molecules. This will change the shape of the Maxwell-Boltzmann distribution. Figure 54 shows a Maxwell-Boltzmann distribution at 300 K and 320 K (the kelvin temperature scale is often used).

By comparing the areas under the curves above each activation energy it becomes clear that at the higher temperature there are many more reactant molecules with enough energy to react. This explains why there is a higher rate of reaction at a higher temperature.

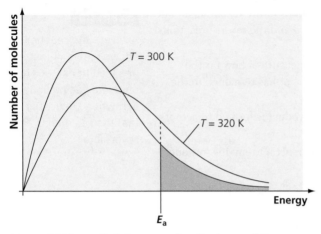

Figure 54 Maxwell-Boltzmann distribution at different temperatures

Exam tip

For Maxwell–Boltzmann distributions at different temperatures, the total area under each graph must be the same, because the concentrations of the reactant molecules are all the same. This means that as the energy of the particles increases the curve becomes lower but more spread out. You will often be asked to sketch a curve like this at a lower or higher temperature. Lower temperature curves are narrower and have a higher peak than higher temperature curves that are wider and have a lower peak.

A small increase in temperature has a greater effect than an increase in concentration. The area under the graph above the activation energy increases to a greater extent for a small increase in temperature than it does for a significant increase in concentration.

Summary

- Activation energy is the minimum energy that the reactant molecules require for a reaction to occur. It can be represented on an enthalpy profile diagram or on a Maxwell–Boltzmann distribution.
- A Maxwell–Boltzmann distribution shows the distribution of molecular energies of the reactant particles. The graph changes shape with changes in temperature.
- A catalyst increases the rate of reaction by providing an alternative reaction route that has a lower activation energy.

Chemical equilibria, Le Chatelier's principle and K_c

Chemical equilibria and Le Chatelier's principle

Reversible reactions go in both the forward and reverse directions. Many reactions are reversible. The reaction starts with the reactants and, as products form, some of these products break down to form reactants again.

- When a reaction system is at equilibrium the concentrations of the reactants and of the products remain constant and the rates of the forward and the reverse reactions are equal.
- The position of equilibrium in a reversible reaction is a measure of how far the reaction has proceeded to the right (towards the products) or has remained to the left (towards the reactants).
  - An equilibrium that lies to the left means that little product is formed (reactants predominate).
  - An equilibrium that lies to the right means that little reactant remains (products predominate).
- Several factors affect the position of equilibrium. These include concentration, temperature and pressure in homogeneous gaseous reactions.
- A homogeneous system is one in which all reactants and products are in the same physical state.

Le Chatelier's principle states that if a factor is changed that affects a system at equilibrium, the position of equilibrium will move in a direction so as to oppose that change.

Changes in concentration

- An increase in the concentration of a reactant moves the position of equilibrium from left to right to oppose the change and remove the added reactant.
- An increase in the concentration of a product moves the position of equilibrium from right to left to oppose the change and remove the added product.
- A decrease in the concentration/removal of a reactant moves the position of equilibrium from right to left to oppose the change and make more of the reactant.
- A decrease in the concentration/removal of a product moves the position of equilibrium from left to right to oppose the change and make more of the product.

Changes in temperature

Changes in temperature affect the position of equilibrium based on whether the reaction is exothermic or endothermic.

If the forward reaction is endothermic (ΔH = +ve):

- An increase in temperature moves the position of equilibrium in the direction of the forward endothermic reaction to absorb the heat. The position of equilibrium moves from left to right (higher yield or higher concentration of products).
- A decrease in temperature moves the position of equilibrium in the direction of the reverse exothermic reaction to release more heat. The position of equilibrium moves from right to left (lower yield or lower concentration of products).

If the forward reaction is exothermic (ΔH = −ve):

- An increase in temperature moves the position of equilibrium in the direction of the reverse endothermic reaction to absorb the heat. The position of equilibrium moves from right to left (lower yield or lower concentration of products).
- A decrease in temperature moves the position of equilibrium in the direction of the forward exothermic reaction to release more heat. The position of equilibrium moves from left to right (higher yield or higher concentration of products).

Exam tip

A reversible reaction is shown with a $\rightleftharpoons$ instead of a traditional $\rightarrow$. The opposite-facing double arrowhead indicates that the reaction is reversible.

Changes in pressure

A change in pressure applies to a homogeneous gas system in which all the reactants and products are gases. Count the total number of moles of gas on the reactant side and on the product side, based on the balanced numbers in the equation. If there is the same number of moles of gas on each side, a change in pressure will have no effect on the position of equilibrium.

If the left-hand side (reactants) has fewer gas moles:

- An increase in pressure moves the position of equilibrium in the direction of the equation side with fewer gas moles. The position of equilibrium moves from right to left (lower yield or lower concentration of products) to oppose the increase in pressure.
- A decrease in pressure moves the position of equilibrium in the direction of the equation side with more gas moles. The position of equilibrium moves from left to right (higher yield or higher concentration of products) to oppose the decrease in pressure.

If the right-hand side (products) has fewer gas moles:

- An increase in pressure moves the position of equilibrium in the direction of the equation side with fewer gas moles. The position of equilibrium moves from left to right (higher yield or higher concentration of products) to oppose the increase in pressure.
- A decrease in pressure moves the position of equilibrium in the direction of the equation side with more gas moles. The position of equilibrium moves from right to left (lower yield or lower concentration of products) to oppose the decrease in pressure.

As a means of increasing product yield, high pressure is expensive as it requires energy to apply using electric pumps and more expensive valves, and it also requires a container, or reaction vessel, which can withstand the increased pressure.

Presence of a catalyst

The presence of a catalyst has no effect on the position of equilibrium but equilibrium is attained more quickly.

Equilibrium in industrial reactions

The Haber process

The industrial production of ammonia from nitrogen and hydrogen is carried out using the Haber process. The reaction is a homogeneous gaseous equilibrium. The equation and enthalpy change of reaction are:

$$N_2 + 3H_2 \rightleftharpoons 2NH_3 \qquad \Delta H^\ominus = -92 \, \text{kJ} \, \text{mol}^{-1}$$

- An increase in temperature causes the position of equilibrium to move from right to left (in the direction of the endothermic process) and there is a lower yield of ammonia.
- An increase in pressure causes the position of equilibrium to move from left to right (fewer gas moles) and there is a higher yield of product.
- Presence of iron catalyst allows equilibrium to be reached more quickly.

> **Exam tip**
>
> It is important to state that the movement of the position of equilibrium occurs to oppose the change applied, as this is the main focus of Le Chatelier's principle.

> **Knowledge check 37**
>
> For the reaction $2HI(g) \rightleftharpoons H_2(g) + I_2(g)$, explain why increasing the pressure has no effect on the position of equilibrium.

The industrial Haber process uses the following reaction conditions: a compromise temperature of 450°C; pressure of 20 MPa, because high pressure increases the yield of NH_3; and an iron catalyst, to attain equilibrium more rapidly.

A lower temperature would give a higher yield of ammonia, but this would also lower the rate of reaction. The temperature used is a compromise between achieving a reasonable yield and a fast enough rate of reaction.

The Contact process

In the Contact process for the production of concentrated sulfuric acid, sulfur dioxide reacts with oxygen to form sulfur trioxide. The equation and enthalpy change of reaction are:

$$SO_2 + \tfrac{1}{2}O_2 \rightleftharpoons SO_3 \qquad \Delta H^\ominus = -98\,kJ\,mol^{-1}$$

- An increase in temperature causes the position of equilibrium to move from right to left (in the direction of the endothermic process) and there is a lower yield of product.
- An increase in pressure causes the position of equilibrium to move from left to right (fewer gas moles) and there is a higher yield of product.
- Presence of vanadium(v) oxide catalyst allows equilibrium to be reached more quickly.

The Contact process reaction conditions used are: a compromise temperature of 450°C; 2 kPa — a slightly higher pressure to increase the yield of SO_3; and a V_2O_5 catalyst to attain equilibrium more rapidly.

Knowledge check 38

Name the catalyst used in the Haber process.

Knowledge check 39

State two features of a reaction at equilibrium.

Summary

- The features of a reaction system at equilibrium are that the concentrations of the reactants and the products remain constant and the rates of the forward and the reverse reactions are equal.
- Le Chatelier's principle states that if a factor is changed that affects a system in equilibrium, the position of equilibrium will move in a direction so as to oppose the change.
- If the concentration of a reactant is increased, the position of equilibrium will move from left to right.

- If the temperature is increased and the forward reaction is exothermic, the position of equilibrium will move from right to left.
- If the pressure is increased, the position of equilibrium will shift to the side with the fewer number of gas moles.
- A catalyst has no effect on the position of equilibrium but will increase the rate of reaction so equilibrium is attained faster.

The equilibrium constant, K_c, for homogeneous systems

K represents the equilibrium constant. The subscript letter after K shows what type of equilibrium is being expressed. K_c is an equilibrium constant calculated from concentrations of reactants and products (in $mol\,dm^{-3}$).

Equilibrium constants are constant only at constant temperature. The temperature should be quoted when the value of any equilibrium constant is given. If temperature remains constant the equilibrium constant will not change. If other factors, such as pressure or concentration of reactants, are changed, then the value of the equilibrium constant will still remain constant as long as temperature remains unchanged.

Exam tip

It is important to remember that the position of a reaction at equilibrium may vary when external factors are changed, but only changes in temperature will affect the value of the equilibrium constant. This is a commonly asked question.

K_c

For the reaction:

$$aA + bB \rightleftharpoons cC + dD$$

$$K_c = \frac{[C]^c[D]^d}{[A]^a[B]^b}$$

where [A] represents the concentration of A in $mol\,dm^{-3}$ in the equilibrium mixture and the superscript 'a' is the balancing number for A in the equation for the reaction. The same applies to B, C and D.

$$\text{units of } K_c = \frac{(mol\,dm^{-3})^{(c+d)}}{(mol\,dm^{-3})^{(a+b)}}$$

The units are in terms of concentration in $mol\,dm^{-3}$ but the overall power depends on the balancing numbers in the equation for the reaction.

Writing K_c expressions and calculation of units

A common question is to write an expression for K_c and to calculate the units of K_c.

> ### Worked example
>
> Write an expression for K_c and calculate its units, for the reaction:
>
> $$2NO(g) + O_2(g) \rightleftharpoons 2NO_2(g)$$
>
> #### Answer
>
> $$K_c = \frac{[NO_2]^2}{[NO]^2[O_2]}$$
>
> $$\text{units of } K_c = \frac{(mol\,dm^{-3})^2}{(mol\,dm^{-3})^3} = mol^{-1}\,dm^3$$

No units of K_c

For the general reaction:

$$A + B \rightleftharpoons C + D$$

$$K_c = \frac{[C][D]}{[A][B]}$$

$$\text{units} = \frac{(mol\,dm^{-3})^2}{(mol\,dm^{-3})^2} = \text{no units}$$

If K_c has no units, then the total volume in the reaction will not make a difference to the calculation of K_c. This means that the number of moles at equilibrium can be used instead of concentration at equilibrium in the K_c expression. This is important if the volume in which the reaction occurs is not given to you.

Calculating K_c and using K_c

You may be required to calculate K_c or to use K_c to calculate an equilibrium number of moles or an initial number of moles or a concentration at equilibrium.

> ### Exam tip
>
> Remember that when the same term is multiplied the powers are added and if the same term is divided then the powers are subtracted. In the worked example, $concentration^2$ divided by concentration = $concentration^{(2-1)}$ = concentration, so the units are $mol\,dm^{-3}$. However, if you have $concentration^2$ divided by $concentration^2$, then that = $concentration^{(2-2)}$ = $concentration^0$ = 1 (no units).

> ### Knowledge check 40
>
> Write a K_c expression for the reaction $N_2O_4(g) \rightleftharpoons 2NO_2(g)$ and state its units.

Content Guidance

Calculating K_c

Follow the general format shown in Figure 55 to calculate number of moles of each substance at equilibrium.

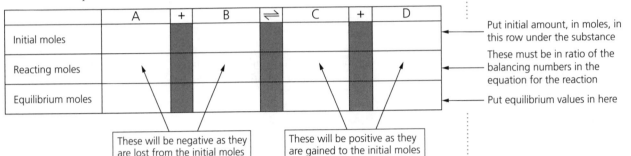

	A	+	B	⇌	C	+	D	
Initial moles								Put initial amount, in moles, in this row under the substance
Reacting moles								These must be in ratio of the balancing numbers in the equation for the reaction
Equilibrium moles								Put equilibrium values in here

These will be negative as they are lost from the initial moles

These will be positive as they are gained to the initial moles

Figure 55

Worked example 1

Calculate a value for K_c for the reaction below at a certain temperature, if 8.0 moles of nitrogen monoxide produce 2.0 moles of nitrogen at equilibrium.

$$2NO(g) \rightleftharpoons N_2(g) + O_2(g)$$

Answer

$$K_c = \frac{[N_2][O_2]}{[NO]^2}$$

$$\text{units of } K_c = \frac{(\text{mol dm}^{-3})^2}{(\text{mol dm}^{-3})^2} = \text{no units}$$

Note that in this question, no volume is given and this is because the equilibrium constant has no units. The total volume would cancel out in the expression so equilibrium moles can be used in place of concentration.

The first step is to fill in the number of moles you are given (Figure 56). Note that as there is only NO present initially, the initial moles of N_2 and O_2 are both zero.

	2NO	⇌	N_2	+	O_2
Initial moles	8.0		0		0
Reacting moles					
Equilibrium moles			2.0		

Figure 56

The second step is to calculate the other number of moles of each substance present at equilibrium (Figure 57). If 2.0 moles of N_2 are formed then the reacting moles of N_2 are +2.0. This means that the reacting moles of NO are −4.0 (based on the ratio in the equation) and the reacting moles of O_2 are +2.0.

	2NO	⇌	N$_2$	+	O$_2$
Initial moles	8.0		0		0
Reacting moles	−4.0		+2.0		+2.0
Equilibrium moles			2.0		

Figure 57

The next step is to complete the equilibrium number of moles of each substance (Figure 58).

For NO it is 8.0 − 4.0 = 4.0 moles

For O$_2$ it is 0 + 2.0 = 2.0 moles

	2NO	⇌	N$_2$	+	O$_2$
Initial moles	8.0		0		0
Reacting moles	−4.0		+2.0		+2.0
Equilibrium moles	4.0		2.0		2.0

Figure 58

As K_c has no units the values of the K_c can be calculated from the equilibrium number of moles.

$$K_c = \frac{[N_2][O_2]}{[NO]^2} = \frac{2.0 \times 2.0}{4.0^2} = 0.25 \text{ no units}$$

If K_c has units, then the equilibrium moles are divided by the volume (which should be given) to calculate the equilibrium concentration of each substance. K_c is then calculated from the equilibrium concentrations in $mol\,dm^{-3}$ at this temperature. If the volume in this question was $10\,dm^3$, the equilibrium concentrations of NO, N$_2$ and O$_2$ are 0.4, 0.2 and 0.2. K_c calculated from these values is:

$$K_c = \frac{0.2 \times 0.2}{0.4^2} = \frac{0.2^2}{0.4^2} = 0.25 \text{ no units}$$

The value is the same no matter what volume is used when K_c has no units.

Calculating K_c from concentrations

The concentrations of the reactants initially may be given as concentrations or number of moles in a specific volume. For a K_c that has units, the number of moles at equilibrium must be in concentration units ($mol\,dm^{-3}$). Carry out the calculation as before, but determine the equilibrium concentrations by dividing the equilibrium moles by the volume (which will be given in the question).

Knowledge check 41

Calculate a value for K_c for the reaction $2A(g) + B(g) \rightleftharpoons C(g)$, if 2 moles of A and 1 mole of B were mixed in a volume of $2\,dm^3$, and 0.4 moles of C were present at equilibrium. State the units of K_c.

Worked example

Nitrogen reacts with hydrogen according to the reaction:

$$N_2(g) + 3H_2(g) \rightleftharpoons 2NH_3(g)$$

0.0425 moles of nitrogen are mixed with 0.0105 moles of hydrogen in a volume of 2.00 dm^3 at 150°C. At equilibrium the number of moles of ammonia is 0.00624 mol. Calculate a value for the equilibrium constant, K_c, at 150°C.

Answer

First, arrange the data that you have been given (Figure 59):

	N_2	+	$3H_2$	$\rightleftharpoons$	$2NH_3$
Initial moles	0.0425		0.0105		0
Reacting moles					
Equilibrium moles					0.00624
Equilibrium concentration					

Figure 59

Next, calculate the number of equilibrium moles (Figure 60):

	N_2	+	$3H_2$	$\rightleftharpoons$	$2NH_3$
Initial moles	0.0425		0.0105		0
Reacting moles	−0.00312		−0.00936		+0.00624
Equilibrium moles	0.03938		0.00114		0.00624
Equilibrium concentration					

Figure 60

Equilibrium concentrations (Figure 61) are calculated by dividing the number of equilibrium moles by the volume (2 dm^3).

	N_2	+	$3H_2$	$\rightleftharpoons$	$2NH_3$
Initial moles	0.0425		0.0105		0
Reacting moles	−0.00312		−0.00936		+0.00624
Equilibrium moles	0.03938		0.00114		0.00624
Equilibrium concentration	0.01969		0.00057		0.00312

Figure 61

$$K_c = \frac{[NH_3]^2}{[N_2][H_2]^3} = \frac{(0.00312)^2}{(0.01969)(0.00057)^3} = \frac{9.7344 \times 10^{-6}}{3.646 \times 10^{-12}} = 2.70 \times 10^6 \, mol^{-2} \, dm^6$$

Exam tip

With calculations of this type there may be rounding differences in the answer achieved, dependent on whether the answer is worked out completely on a calculator, or whether it has been rounded to 3 or 4 significant figures at different stages of the calculation. Either final answer is valid as long as all working is shown.

Percentage conversion questions

In some questions about K_c you may not be told the actual amounts of the substances. Instead, you might be told the relative amounts (and what percentage of a substance is converted). In such cases you must choose suitable amounts to consider.

For example, if you are told that you are starting with an equimolar mixture of X and Y you will probably consider the case of 1 mole of X and 1 mole of Y. Likewise if the initial mixture has twice as many moles of Y as of X, you will probably choose 1 mole of X and 2 moles of Y, although 2 and 4 moles, or 3 and 6 moles etc., would also do. If you are told the position of equilibrium in terms of a percentage, be careful to note whether it is the percentage of the reactant that has been converted or the percentage that remains unconverted.

Worked example

The following equilibrium is established at 444°C when 1.00 mole of HI reacts.

$$2HI(g) \rightleftharpoons H_2(g) + I_2(g)$$

It is found that the HI has undergone 22% dissociation.

Write an expression for the equilibrium constant, K_c, and use it to calculate a value of the equilibrium constant, K_c, at 444°C, and state its units.

Answer

$$K_c = \frac{[HI]^2}{[H_2][I_2]}$$

22% dissociation means that 22% of the 1 mole of HI reacts. So 0.22 mole of HI reacts. There are no units of this K_c value, so the volume is not required and equilibrium moles can be used to calculate K_c (Figure 62).

	2HI	$\rightleftharpoons$	H$_2$	+	I$_2$
Initial moles	1		0		0
Reacting moles	−0.22		+0.11		+0.11
Equilibrium moles	0.78		0.11		0.11

Figure 62

$$K_c = \frac{[H_2][I_2]}{[HI]^2} = \frac{(0.11)(0.11)}{(0.78)^2} = 0.0199 \qquad \text{no units}$$

Knowledge check 42

Calculate a value for K_c for the reaction $A(g) + B(g) \rightleftharpoons 2C(g)$, when the concentrations of A, B and C at equilibrium are $0.0152\,mol\,dm^{-3}$, $0.0174\,mol\,dm^{-3}$ and $0.0244\,mol\,dm^{-3}$, respectively.

Calculations using K_c

You may be asked to use a K_c value to determine a concentration or number of moles of a reactant or product at equilibrium.

These types of calculation simply require the rearrangement of the K_c expression to calculate a concentration.

For the reaction

$$A(g) + 2B(g) \rightleftharpoons 2C(g)$$

4.00 mol of A and 1.50 mol of B were found to be present in a container of volume 2.5 dm^3 at equilibrium, at 250°C. K_c for this reaction at 250°C is 0.174 mol^{-1} dm^3. Calculate the number of moles of C present in the 2.5 dm^3 container.

Answer

$$\text{concentration of A} = \frac{4.00}{2.5} = 1.6 \, \text{mol dm}^{-3}$$

$$\text{concentration of B} = \frac{1.50}{2.5} = 0.6 \, \text{mol dm}^{-3}$$

$$K_c = \frac{[C]^2}{[A][B]^2}$$

$$[C]^2 = K_c[A][B]^2$$

$$[C]^2 = 0.174 \times 1.6 \times (0.6)^2$$

$$[C]^2 = 0.1002$$

$$[C] = \sqrt{0.1002} = 0.3165 \, \text{mol dm}^{-3}$$

$$\text{moles of C in 2.5 dm}^3 = 0.3165 \times 2.5 = 0.791 \, \text{mol dm}^{-3}$$

When K_c has units, the values used for concentration must be in mol dm^{-3} (that's why we divided by 2.5 initially in this worked example). At the end, it is the number of moles in 2.5 dm^3 so the final concentration of C is × 2.5 to convert back to the 2.5 dm^3 volume.

■ Oxidation, reduction and redox equations

Redox is oxidation and reduction occurring simultaneously in the same reaction.

There are four different definitions of both oxidation and reduction, as shown in Table 20.

Oxidation	Reduction
Loss of electrons	Gain of electrons
Gain of oxygen	Loss of oxygen
Loss of hydrogen	Gain of hydrogen
Increase in oxidation state	Decrease in oxidation state

Table 20

Oxidation state is the charge on a simple ion, or the difference in the number of electrons associated with an atom in a compound compared with the atoms of the element.

The **oxidation state** in a compound is defined as the hypothetical charge on an atom assuming that the bonding is completely ionic.

Working with oxidation states

There are rules for working with oxidation states.

1 The oxidation state of the atoms in an element is 0 (zero). For example:
   – the oxidation state of Na atoms in sodium metal is 0

- the oxidation state of both Cl atoms in Cl_2 is 0
- the oxidation state of all eight S atoms in S_8 is 0

2 Non-zero oxidation states are always written as positive or negative integers (whole numbers), for example +2, –1, +7.

3 Fractional oxidation states are possible, but only as the average of several atoms in a compound. For example, in Fe_3O_4 the oxidation state of iron is $+2\frac{2}{3}$ as two of the iron atoms have an oxidation state of +3 and one has an oxidation state of +2 so the average is $+2\frac{2}{3}$.

4 Oxygen has an oxidation state of –2 in almost all compounds, except in peroxides (for example, H_2O_2, where it is –1) and in the compound OF_2, where it is +2 (fluorine is more electronegative than oxygen).

5 Hydrogen has an oxidation state of +1 in almost all compounds except hydrides, for example NaH, where it is –1.

6 Group 1 elements have an oxidation state of +1 in all compounds.

7 Group 2 elements have an oxidation state of +2 in all compounds.

8 The oxidation state of ions in a compound is equal to the charge on the ion. For example:
- in iron(II) chloride, iron has an oxidation state of +2
- in copper(II) sulfate, copper has an oxidation state of +2
- in silver(I) nitrate, silver has an oxidation state of +1
- in sodium chloride, chlorine has an oxidation state of –1
- in magnesium oxide, oxygen has an oxidation state of –2

9 The total of the oxidation states for the atoms in a compound must equal 0 (zero).

10 The total of the oxidation states of the atoms in a molecular ion must equal the charge on the ion. For example:
- the total of the oxidation states of the atoms of the elements in sulfate, SO_4^{2-}, must equal –2
- the total of the oxidation states of the atoms of the elements in nitrate, NO_3^-, must equal –1

11 The oxidation states of the p and d block elements vary significantly.

12 The maximum oxidation state of a p block element is '+ group number'. For example:
- the maximum oxidation state of Cl is +7
- the maximum oxidation state of N is +5

13 The minimum oxidation state of a p block element is 'group number – (minus) 8'. For example:
- the minimum oxidation state of Cl = 7 – 8 = –1
- the minimum oxidation state of N = 5 – 8 = –3

14 The oxidation states of d block elements can vary up to +7.

Calculating oxidation state

The rules described for working with oxidation states are used to calculate oxidation states of different atoms in compounds and ions.

Worked example 1

Determine the oxidation state of S in sodium sulfate, Na_2SO_4.

Answer

Na: oxidation state = +1; two Na atoms present, so total for 2 Na = +2

S: oxidation state = x (the unknown)

O: oxidation state = −2; four O atoms present, so total for 4 O = −8

Na_2SO_4: $+2 + x − 8 = 0$ (the total of the oxidations state is zero, as it is a compound)

$x = +8 − 2 = +6$

oxidation state of S in Na_2SO_4 = +6

Worked example 2

Determine the oxidation state of Cr in the dichromate ion, $Cr_2O_7^{2-}$.

Cr: oxidation state = x; two Cr atoms present, so $2x$

O: oxidation state = −2; seven O atoms present, so total = −14

$Cr_2O_7^{2-}$: $2x − 14 = −2$ (total of oxidation state is −2 as it is an ion)

$2x = −2 + 14 = +12$

$x = +6$

oxidation state of each Cr atom in $Cr_2O_7^{2-}$ = +6

> **Knowledge check 43**
>
> Calculate the oxidation state of nitrogen in HNO_2.

Explaining redox

Remember it is often elements from the middle or right-hand side of the periodic table that are oxidised and reduced, but watch out for elements from other groups that are oxidised from/reduced to a zero oxidation state.

When given a redox equation and asked to explain why it is described as redox, you need to calculate the oxidation states of the elements that are oxidised or reduced. Figure 63 gives an example that shows how to do this.

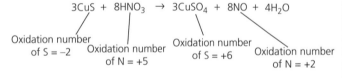

$$3CuS + 8HNO_3 \rightarrow 3CuSO_4 + 8NO + 4H_2O$$

Oxidation number of S = −2   Oxidation number of N = +5   Oxidation number of S = +6   Oxidation number of N = +2

- Sulfur is oxidised from −2 to +6
- Nitrogen is reduced from +5 to +2
- Redox is where oxidation and reduction occur in the same reaction

Figure 63 Explaining redox

Proper names for compounds

In sodium sulfate the oxidation state of the sulfur is +6, so the proper name for sodium sulfate is sodium sulfate(VI). The VI represents the oxidation state of the S in sulfate.

In the dichromate ion, $Cr_2O_7^{2-}$, the oxidation state of both chromium atoms is +6, so the proper name for the dichromate ion is dichromate(VI) ion.

Half-equations

A half-equation is an oxidation or reduction equation involving loss or gain of electrons. Examples of simple half-equations:

$$Mg \rightarrow Mg^{2+} + 2e^-$$

This is an oxidation, as 1 mole of magnesium atoms loses 2 moles of electrons to form 1 mole of magnesium ions.

$$Cl_2 + 2e^- \rightarrow 2Cl^-$$

This is a reduction, as 1 mole of chlorine molecules gains 2 moles of electrons to form 2 moles of chloride ions.

$$Fe^{2+} \rightarrow Fe^{3+} + e^-$$

This is an oxidation, as 1 mole of iron(ɪɪ) ions loses 1 mole of electrons to form 1 mole of iron(ɪɪɪ) ions.

More complex half-equations involve calculation of oxidation states and balancing any oxygen atoms gained or lost using H^+ ions and H_2O.

A quick way is to balance the atoms changing oxidation state, put in H^+ ions and H_2O to balance any change in oxygen.

Exam tip

If the half-equation is oxidation, the electrons are on the right-hand side. If the half-equation is reduction the electrons are on the left-hand side.

Worked example 1

Manganate(ᴠɪɪ), MnO_4^-, can be reduced to manganese(ɪɪ), Mn^{2+} (Figure 64).

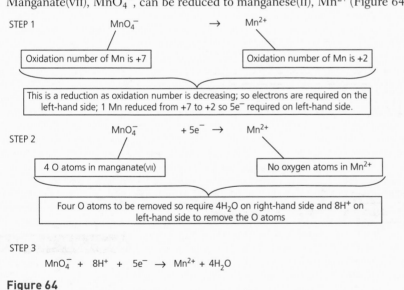

Figure 64

Exam tip

The overall charge on the left and on the right of any half-equation or ionic equation should be the same if the equation is balanced correctly. Check that the overall charges on both sides match, to make sure that the half-equation is correct.

Left total charge = $-1 + 8 - 5 = +2$
Right total charge = $+2$

Worked example 2

Dichromate(VI), $Cr_2O_7^{2-}$ can be reduced to chromium(III), Cr^{3+} (Figure 65).

STEP 1 $Cr_2O_7^{2-}$ $\rightarrow$ Cr^{3+}

Oxidation number of Cr is +6   Oxidation number of Cr is +3

This is a reduction as oxidation number is decreasing; so electrons are required on the left-hand side; 2 Cr reduced from +6 to +3 so 6e⁻ required on left-hand side; also need a 2 in front of Cr^{3+}

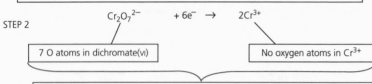

STEP 2 $Cr_2O_7^{2-}$ $+ 6e^-$ $\rightarrow$ $2Cr^{3+}$

7 O atoms in dichromate(VI)   No oxygen atoms in Cr^{3+}

Seven O atoms to be removed so require $7H_2O$ on right-hand side and $14H^+$ on left-hand side to remove the O atoms

STEP 3

$$Cr_2O_7^{2-} + 14H^+ + 6e^- \rightarrow 2Cr^{3+} + 7H_2O$$

Figure 65

Worked example 3

Oxidation of sulfur dioxide, SO_2 to sulfate, SO_4^{2-} (Figure 66)

STEP 1 SO_2 $\rightarrow$ SO_4^{2-}

Oxidation number of S is +4   Oxidation number of S is +6

This is an oxidation as oxidation number is increasing; so electrons are required on the right-hand side; 1 S oxidised from +4 to +6 so 2e⁻ required on right-hand side.

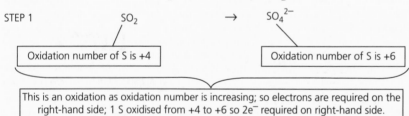

STEP 2 SO_2 $\rightarrow$ SO_4^{2-} $+$ $2e^-$

2 O atoms in SO_2 4 O atoms in sulfate

Two O atoms to be added so $2H_2O$ added to left-hand side and $4H^+$ on right-hand side

STEP 3

$$SO_2 + 2H_2O \rightarrow SO_4^{2-} + 2e^- + 4H^+$$

Figure 66

Knowledge check 44

Write a half-equation for the reduction of sulfate(VI) ions to sulfur.

Balancing redox equations

Half-equations include electrons. A half-equation is half a redox reaction and involves the oxidation or reduction of one particular species. Examples of half-equations are:

$$Ni \rightarrow Ni^{2+} + 2e^-$$

$$Cr_2O_7^{2-} + 14H^+ + 6e^- \rightarrow 2Cr^{3+} + 7H_2O$$

Features of half-equations:

■ they involve electrons
■ only one species is oxidised or reduced

Ionic equations or redox equations do not include electrons. An ionic equation is the reaction between two ionic species transferring electrons. Examples of ionic equations are:

$$Mg + 2H^+ \rightarrow Mg^{2+} + H_2$$

$$Cl_2 + 2I^- \rightarrow 2Cl^- + I_2$$

Features of ionic equations:

■ they do not involve electrons
■ two species involved — one oxidised, one reduced

Often, two half-equations are given to you and you are asked to write the ionic equation. This is simply a matter of multiplying the half-equations by a number that gives the same number of electrons in the oxidation half-equation and in the reduction half-equation.

When the equations are added together to make an ionic equation, there will be the same number of electrons on both sides of the ionic equation so they can be cancelled out.

There are two ways in which half-equations may be presented:

1　You may be given one half-equation that is a reduction and the other will be written as an oxidation.
2　You may be given two half-equations that are both written as reductions. In this example one equation needs to be reversed to make it an oxidation before you can add the equations together.

Adding half-equations together

Make sure you have one oxidation equation (electrons on the right) and one reduction (electrons on the left). For example:

$$Al \rightarrow Al^{3+} + 3e^- \qquad \text{Oxidation}$$

$$F_2 + 2e^- \rightarrow 2F^- \qquad \text{Reduction}$$

Make sure that the numbers of electrons are the same in both the oxidation and reduction half-equations. To do this the oxidation equation needs to be multiplied by 2. The reduction equation needs to be multiplied by 3. This will give both equations six electrons:

$$2Al \rightarrow 2Al^{3+} + 6e^-$$

$$3F_2 + 6e^- \rightarrow 6F$$

To add them, simply write down all the species from the left-hand side of both half-equations, then put an arrow and finally write down all the species from the right-hand side of both half-equations:

$$Al + 3F_2 + 6e^- \rightarrow 2Al^{3+} + 6e^- + 6F^-$$

The next step is to cancel out the electrons on both sides of the ionic equation:

$$Al + 3F_2 \rightarrow 2Al^{3+} + 6F^-$$

This is the ionic equation for the reaction between aluminium and fluorine.

> **Exam tip**
>
> Always check ionic equations for charges on the left-hand side and the right-hand side — the total charges should be the same on both sides: 0 on the left-hand side and 0 on the right-hand side.

Worked example 1

Iron(II) ions are oxidised by acidified potassium manganate(VII). Write an equation for the reaction between iron(II) ions and manganate(VII) ion.

The two half-equations are:

oxidation: $Fe^{2+} \rightarrow Fe^{3+} + e^-$

reduction: $MnO_4^- + 8H^+ + 5e^- \rightarrow Mn^{2+} + 4H_2O$

Answer

Note that there are five electrons on the left-hand side of the second equation and only one electron on the right-hand side of the first equation.

In order to write a complete ionic equation the first equation must be multiplied by 5, and then the two equations are added together:

$$5Fe^{2+} \rightarrow 5Fe^{3+} + 5e^-$$

$$MnO_4^- + 8H^+ + 5e^- \rightarrow Mn^{2+} + 4H_2O$$

$$\overline{MnO_4^- + 8H^+ + 5e^- + 5Fe^{2+} \rightarrow Mn^{2+} + 4H_2O + 5Fe^{3+} + 5e^-}$$

The $5e^-$ on each side can be cancelled, so that the overall equation reads:

$$MnO_4^- + 8H^+ + 5Fe^{2+} \rightarrow Mn^{2+} + 4H_2O + 5Fe^{3+}$$

Sometimes the half-equations are given as two reductions, so one equation must be reversed to enable the electrons to be eliminated. The reaction will indicate which two species are reacting. The two half-equations in worked example 1 could have been presented as:

reduction: $Fe^{3+} + e^- \rightarrow Fe^{2+}$

reduction: $MnO_4^- + 8H^+ + 5e^- \rightarrow Mn^{2+} + 4H_2O$

You are asked to write an ionic equation for the reaction between iron(II) ions, Fe^{2+}, and manganate(VII) ions, MnO_4^-. The reaction requires the first half-equation to be reversed to $Fe^{2+} \rightarrow Fe^{3+} + e^-$ and then multiplied by 5 as before.

The final equation is the same when the electrons have been eliminated.

$$MnO_4^- + 8H^+ + 5Fe^{2+} \rightarrow Mn^{2+} + 4H_2O + 5Fe^{3+}$$

Worked example 2

Write an ionic equation for the reaction of nitrate(III) ions, NO_2^-, and dichromate(VI) ions, $Cr_2O_7^{2-}$, using the following half-equations:

$$NO_2^- + H_2O \rightarrow NO_3^- + 2H^+ + 2e^-$$

$$Cr_2O_7^{2-} + 14H^+ + 6e^- \rightarrow 2Cr^{3+} + 7H_2O$$

Answer

The equations given are an oxidation (first equation) and a reduction (second equation) so that they can be combined directly once the electrons have been balanced:

$$3NO_2^- + 3H_2O \rightarrow 3NO_2^- + 6H^+ + 6e^-$$

$$Cr_2O_7^{2-} + 14H^+ + 6e^- \rightarrow 2Cr^{3+} + 7H_2O$$

Add the left-hand sides and right-hand sides together:

$$Cr_2O_7^{2-} + 14H^+ + 3NO_2^- + 3H_2O + 6e^- \rightarrow 2Cr^{3+} + 7H_2O + 3NO_3^- + 6H^+ + 6e^-$$

The electrons, water and H^+ can be cancelled to remove them from both sides of the equation:

$$Cr_2O_7^{2-} + 8H^+ + 3NO_2^- \rightarrow 2Cr^{3+} + 4H_2O + 3NO_3^-$$

Knowledge check 45

Explain why the reaction

$$H_2O_2 + 2KI + H_2SO_4 \rightarrow I_2 + K_2SO_4 + 2H_2O$$

is described as a redox reaction, and write a half-equation for the reduction of H_2O_2.

Summary

- The oxidation number (or oxidation state) of a particular atom in a compound or ion is defined as the hypothetical charge on an atom assuming that the bonding is completely ionic.
- Hydrogen has an oxidation state of +1 in almost all compounds and ions except hydrides.
- Oxygen has an oxidation state of –2 in almost all compounds and ions except peroxides.
- Group 1 atoms in a compound have an oxidation state of +1; group 2 atoms +2; and Al +3.
- The oxidation states of transition atoms in compounds and ions vary.
- Oxidation states should always be stated with a + or – sign.
- The total oxidation state in a compound equals zero.
- The total oxidation state in a molecular ion equals the charge on the ion.
- Half-equations represent oxidation and reduction; reduction is gain of electrons and oxidation is loss of electrons.
- When combining half-equations the equations are multiplied so that the number of electrons cancel out, then any H^+ ions and H_2O on either side can cancel out as well.

Questions & Answers

This section contains a mix of multiple-choice and structured questions similar to those you can expect to find in the AS and A-level papers.

Papers 1 and 2 of the AS examinations consist of 15 multiple-choice questions (each with four options, A to D), followed by 65 marks of structured questions, totalling 80 marks. The papers are both 1½ hours. Papers 1 and 2 of the A-level consist of short and long-answer questions (some of which are on the topics covered in this book) totalling 105 marks. Paper 3 of the A-level is a 2-hour paper worth 90 marks. It contains questions worth 40 marks on practical techniques and data analysis, of which there are examples in this guide. There are also questions worth 50 marks, of which 30 marks are multiple-choice questions, testing across the whole specification. For both the AS and A-level papers, 15% of the marks cover practical aspects of the specification and 20% mathematical content.

About this section

Each question in this section is followed by brief guidance on how to approach the question and also where you could make errors (shown by the icon ⓔ). The answers given are those that examiners would expect from a top-grade candidate. Answers are followed by comments that explain why the answers are correct and indicate pitfalls to avoid. These are preceded by the icon ⓔ. You could try the questions first to see how you get on and then check the answers and comments.

General tips

- Be accurate with your learning at this level as examiners will penalise incorrect wording.
- At both AS and A-level, at least 20% of the marks in assessments for chemistry will require the use of mathematical skills. For any calculation, always follow it through to the end even if you feel that you have made a mistake, as there may be marks for the correct method even if the final answer is incorrect.
- Always attempt to answer multiple-choice questions (there are 15 marks available on both papers 1 and 2 at AS and 30 marks on paper 3 at A-level for multiple-choice questions) even if it is a guess —you have a 25% chance of getting it right.

Atomic structure
Question 1

Approximately how many electrons would have the same mass as the mass of one neutron?

A 20   **B** 200   **C** 2000   **D** 20 000

Answer is C ✓

(e) This question relies on the fact that you know that one electron has a relative mass of 1/1840 that of a proton or a neutron. This means that 1840 electrons would have the mass of one neutron.

Question 2

Which of the following is the electron configuration of a titanium ion, Ti^{2+}?

A $1s^2\, 2s^2\, 2p^6\, 3s^2\, 3p^6\, 4s^2$

B $1s^2\, 2s^2\, 2p^6\, 3s^2\, 3p^6\, 3d^2\, 4s^2$

C $1s^2\, 2s^2\, 2p^6\, 3s^2\, 3p^6\, 3d^4$

D $1s^2\, 2s^2\, 2p^6\, 3s^2\, 3p^6\, 3d^2$

> Answer is D ✓

(e) The atomic number of titanium is 22 so a titanium atom has 22 protons and 22 electrons. The electron configuration of a titanium atom is $1s^2\, 2s^2\, 2p^6\, 3s^2\, 3p^6\, 3d^2\, 4s^2$. The Ti^{2+} ion has lost 2 electrons. Remember that transition metal atoms lose their 4s electrons first. Hence Ti^{2+} is $1s^2\, 2s^2\, 2p^6\, 3s^2\, 3p^6\, 3d^2$ — the $4s^2$ electrons were lost.

Question 3

The mass spectrum of zirconium (atomic number 40) indicates five different isotopes with the relative abundances shown in Table 1.

Relative isotopic mass	Relative abundance
90	51.5
91	11.2
92	17.1
94	17.4
96	2.8

Table 1

Calculate the relative atomic mass of zirconium to one decimal place. (3 marks)

$$\text{relative atomic mass} = \frac{(90 \times 51.5) + (91 \times 11.2) + (92 \times 17.1) + (94 \times 17.4) + (96 \times 2.8)}{51.5 + 11.2 + 17.1 + 17.4 + 2.8}$$

$$= \frac{9131.8}{100} \quad ✓✓$$

$$= 91.318 = 91.3 \text{ to 1 decimal place } ✓$$

(e) When calculating relative atomic mass from relative isotopic mass data or from a mass spectrum directly, you should multiply each mass by the relative abundance for each isotope, then add them all up. Finally, divide by the total of all the relative abundances. If the question asks for a specific number of decimal places, stick to this or you will lose a mark.

Question 4

(a) Write an equation, including state symbols, to show the process that occurs when the first ionisation energy of calcium is measured. (1 mark)

(b) Explain why the first ionisation energy of rubidium is larger than the first ionisation energy of potassium. (2 marks)

(a) $K(g) \rightarrow K^+(g) + e^-$ ✓

ⓔ 1 mark is awarded for the correct equation and the correct state symbols. To answer this, remember the definition of first ionisation energy — gaseous atoms losing 1 mole of electrons and forming gaseous 1+ ions. The most common error in this answer would be the incorrect state symbols for the potassium atom and the ion. Electrons are lost so should appear on the right-hand side of the equation.

(b) Outer electron for rubidium is further from the nucleus/atomic radius increases ✓.

Outer electron is shielded by more inner shells of electrons ✓.

ⓔ First you need to establish that both potassium and rubidium are in group 1 and that both have outer shell s^1 electron configurations, so there is no difference in stability of filled or half filled subshells. Rubidium is further down group 1 so it has a greater atomic radius, and so more shells of electrons to shield the outer electron from the nuclear charge. As a result there is less nuclear attraction for the outer electron, and so less energy is needed to remove it. Again, knowledge of the definition of ionisation energy is essential.

Amount of substance

Question 1

Which of the following contains the greatest number of atoms?

A   32.1 g of sulfur molecules, S_8

B   24.3 g of magnesium atoms, Mg

C   4.0 g of hydrogen molecules, H_2

D   31.0 g of phosphorus molecules, P_4

Answer is C ✓

ⓔ In this question you can easily calculate the number of particles. Calculate the number of moles using the RFM of the formula given and then multiply by N_A to work out the number of atoms for C and the number of molecules for A, B and D. The number of atoms for A, B and D can be determined by multiplying the number of molecules by the number of atoms in each molecule.

A $\dfrac{mass(g)}{M_r} = \dfrac{32.1}{256.8} = 0.125\,mol \times L\ (6.02 \times 10^{23}) = 7.525 \times 10^{22}$ molecules of S_8

Each S_8 contains eight sulfur atoms so number of atoms $= 8 \times 7.525 \times 10^{22}$

$= 6.02 \times 10^{23}$ atoms of S

B $\dfrac{mass(g)}{M_r} = \dfrac{24.3}{24.3} = 1\,mol \times L\ (6.02 \times 10^{23}) = 6.02 \times 10^{23}$ atoms of Mg

C $\dfrac{mass(g)}{M_r} = \dfrac{4.0}{2.0} = 2\,mol \times L\ (6.02 \times 10^{23}) = 1.204 \times 10^{24}$ molecules of H_2

Each H_2 contains two hydrogen atoms so number of atoms $= 2 \times 1.204 \times 10^{24}$

$= 2.408 \times 10^{24}$ atoms of H

D $\dfrac{mass(g)}{M_r} = \dfrac{31.0}{124.0} = 0.25\,mol \times L\ (6.02 \times 10^{23}) = 1.505 \times 10^{23}$ molecules of P

Each P_4 contains four phosphorus atoms, so

number of atoms $= 4 \times 1.505 \times 10^{23}$

$= 6.02 \times 10^{23}$ atoms of P

Question 2

1.27 g of a hydrated sample of copper(II) sulfate, $CuSO_4.xH_2O$ were heated to constant mass. The mass reduced by 0.46 g. Which of the following is the value of x?

A 3   **B** 4   **C** 5   **D** 6

Answer is C ✓

ⓔ The masses that need to be determined in this style of calculation are the mass of the anhydrous copper(II) sulfate and the mass of water lost. Read the question carefully. The decrease in mass on heating to constant mass is the mass of water lost $(= 0.46\,g)$. The mass of the anhydrous solid, $CuSO_4$, is the mass remaining after heating to constant mass $(= 1.27 - 0.46 = 0.81\,g)$. The moles of anhydrous solid and water can be determined and the simplest ratio worked out.

moles of anhydrous solid, $CuSO_4 = \dfrac{0.81}{159.6} = 0.00508\,mol$

moles of water $= \dfrac{0.46}{18.0} = 0.0256\,mol$

simplest ratio of $CuSO_4 : H_2O = 1 : 5$

so $x = 5$

Or alternatively, the number of moles of anhydrous solid (0.00508) are the same as the number of moles of the hydrated compound, so using the mass 1.27 g, the M_r is $1.27/0.00508 = 250$. Subtract the mass of $CuSO_4$ $(= 158.7)$ leaves

(250 − 158.7) = 91.3. Divide this by the M_r of water (18.0) and the answer is 5.07, so $x = 5$. Rounding of answers when calculating moles may not give an exact whole number but it should be very close.

Question 3

Which one of the titration–indicator pairings in Table 2 would be most suitable to detect the end point of the given acid–base titration?

	Titration	Indicator
A	Ethanoic acid and sodium hydroxide solution	Methyl orange
B	Nitric acid and ammonia solution	Phenolphthalein
C	Sodium hydroxide solution and sulfuric acid	Phenolphthalein
D	Ammonia solution and ethanoic acid	Methyl orange

Table 2

Answer is C ✓

ⓔ The choice of an indicator in a titration depends on the acid and alkali involved. Strong acids (hydrochloric acid, sulfuric acid and nitric acid) reacting with a strong alkali (sodium hydroxide solution or potassium hydroxide solution) can use either phenolphthalein or methyl orange. A strong acid with a weak alkali (usually ammonia solution or sodium carbonate solution) must use methyl orange. A weak acid (mostly organic acids like ethanoic acid) with a strong alkali must use phenolphthalein. There is no suitable indicator for a weak acid with a weak alkali, and a pH meter is the most suitable way of monitoring such a titration.

Question 4

What volume and concentration of sodium hydroxide solution is 20.0 cm³ of 0.03 mol dm⁻³ sulfuric acid exactly neutralised by?

A 15.0 cm³ of 0.02 mol dm⁻³ sodium hydroxide solution

B 15.0 cm³ of 0.04 mol dm⁻³ sodium hydroxide solution

C 30.0 cm³ of 0.02 mol dm⁻³ sodium hydroxide solution

D 30.0 cm³ of 0.04 mol dm⁻³ sodium hydroxide solution

ⓔ The most common error in this question would be to assume a 1 : 1 ratio in the reaction between sodium hydroxide and sulfuric acid. Always write the balanced symbol equation for the titration reaction to check the ratio.

$$2NaOH + H_2SO_4 \rightarrow Na_2SO_4 + 2H_2O$$

The ratio of NaOH to H_2SO_4 is 2 : 1. 20.0 cm³ of 0.03 mol dm⁻³ sulfuric acid is 0.0006 mol. This reacts with 0.0012 mol of NaOH. Which of the answers A to D gives 0.0012 mol of NaOH?

Answer is D ✓

Question 5

0.715 g of a sample of hydrated sodium carbonate, $Na_2CO_3.xH_2O$, were dissolved in deionised water and the volume of the solution was made up to 250 cm³ in a volumetric flask. 25.0 cm³ of this sample was titrated against 0.0500 mol dm⁻³ hydrochloric acid using methyl orange indicator. The average titre was 12.30 cm³.

$$Na_2CO_3 + 2HCl \rightarrow 2NaCl + CO_2 + H_2O$$

(a) State the colour change observed at the end point.   (1 mark)

(b) Calculate the amount, in moles, of hydrochloric acid used in this titration.   (1 mark)

(c) Calculate the amount, in moles, of sodium carbonate present in 25.0 cm³ of solution.   (1 mark)

(d) Calculate the amount, in moles, of sodium carbonate present in 250 cm³ of solution.   (1 mark)

(e) Calculate the M_r of the hydrated sodium carbonate, $Na_2CO_3.xH_2O$.   (1 mark)

(f) Calculate the value of x in $Na_2CO_3.xH_2O$.   (2 marks)

e This is a standard type of structured titration question. The question leads you through the answer and guides you to determine the moles of the anhydrous salt and the moles of water, and then to find the value of x (degree of hydration) by determining the simplest ratio of anhydrous salt to water.

Remember when determining the colour change at the end point that the acid is added to the alkali. This is clear as the average titre is for hydrochloric acid added from the burette.

(a) yellow to red ✓

---

e Even if you cannot work out which solution is being added from the burette, make an educated guess at the colour change for methyl orange as it is always either red to yellow or yellow to red.

(b) moles of HCl = $\dfrac{\text{solution volume } (cm^3) \times \text{concentration } (mol\,dm^{-3})}{1000}$

$= \dfrac{12.3 \times 0.0500}{1000} = 0.000615\,mol$ ✓

---

e The first step in most titration questions is to calculate the number of moles of the solute in the solution added from the burette. Make sure that you do not confuse the solutions volumes as 25.0 cm³ was added to the conical flask and 12.30 cm³ was added from the burette. Check twice that the volume you are using in the calculation is from the correct solution. The average titre should be given to two decimal places.

Questions & Answers

(c) Ratio of $Na_2CO_3 : HCl$ is $1:2$,

so the moles of $Na_2CO_3 = \dfrac{\text{moles of HCl}}{2}$

$= \dfrac{0.000615}{2} = 0.0003075$ ✓

ⓔ The second step in a titration is usually working out the number of moles of solute dissolved in the solution volume (usually $25.0\,cm^3$) that was pipetted into the conical flask. The ratio of the reaction between the two solutes allows you to do this. 1 mole of Na_2CO_3 reacts with 2 moles of HCl so to move from HCl to Na_2CO_3 you need to divide the number of moles by 2. Think this step out logically each time, as errors are often made here with candidates multiplying by 2 instead of dividing by 2.

(d) $0.0003075 \times 10 = 0.003075$ ✓

ⓔ The number of moles of this solute (Na_2CO_3) in $25.0\,cm^3$ of solution is multiplied by 10 to determine the number of moles of the solute in $250\,cm^3$ of the solution, as a $25.0\,cm^3$ sample of the solution was taken, so one tenth of the moles were taken from the volumetric flask. This is the number of moles of Na_2CO_3 in solution and is the same as the number of moles of solid $Na_2CO_3.xH_2O$ that were added to the solution. It is important to realise the distinction between solid $Na_2CO_3.xH_2O$ that is hydrated and sodium carbonate in solution, which is simple $Na_2CO_3(aq)$. The water of crystallisation is not part of the mass of the solute in solution, but the number of moles of each is the same.

(e) $\dfrac{0.715}{0.003075} = 232.52$ ✓

ⓔ The mass of $Na_2CO_3.xH_2O$ divided by the moles gives the M_r. Remember that the moles of Na_2CO_3 in $250\,cm^3$ is exactly the same as the moles of $Na_2CO_3.xH_2O$.

(f) $232.52 - 106.0 = 126.5$ ✓ $\dfrac{126.5}{18.0} = 7.02$ $x = 7$ ✓

ⓔ Subtracting the mass of the anhydrous Na_2CO_3 (106.0) from the M_r of the hydrated $Na_2CO_3.xH_2O$ leaves only the mass of the water (126.5). Dividing this by the M_r of water (18.0) will give the values of x.

Rounding errors during the calculation and also during the titration may have resulted in a rough answer, so the answer provided should be the nearest whole number; $x = 7$. Note that x does not have to be a whole number as hydrated salts lose water of crystallisation gradually over time, so the actual value may be between 6 and 7, although but this would be indicated in the question.

Question 6

For the following reaction:

$V_2O_5(s) + 5Ca(s) \rightarrow 2V(l) + 5CaO(s)$

(a) Calculate the percentage atom economy for the reaction in which vanadium is the useful product. Give your answer to 3 significant figures. (3 marks)

(b) 2.28 kg of vanadium were obtained from 5.00 kg of vanadium(v) oxide in this reaction, where calcium was in excess. Calculate the percentage yield. Give your answer to 3 significant figures. (4 marks)

(a) molecular mass of desired product 2V = 101.8 ✓

sum of molecular masses of all reactants = $V_2O_5 + 5Ca = 382.3$ ✓

% atom economy = $\dfrac{M_r \text{ of desired product}}{\text{sum of } M_r \text{ of all reactants}} \times 100 = \dfrac{101.8}{382.3} \times 100 = 26.6\%$ ✓

ⓔ The most common error here would be to not include the balancing numbers. Atom economies must take into account all atoms in the reaction — two V atoms out of two V, five O and five Ca atoms are desirable or useful in this reaction. This gives a low percentage atom economy. If a use could be found for the CaO produced, the reaction would not be wasteful.

(b) moles of $V_2O_5 = \dfrac{5000}{181.8} = 27.5 \text{ mol}$ ✓

theoretical yield of V = 55.0 mol ✓ (1 : 2 ratio of V_2O_5 : V)

theoretical yield of V(g) = 55.0 × 50.9 = 2799.78 g ✓

% yield = $\dfrac{2280}{2799.78} \times 100 = 81.4\%$ ✓

ⓔ This calculation is carried out using masses to determine the percentage yield but equally well moles could be used. The theoretical yield of V in moles is 55.00 mol. The mass of V obtained is 2.28 kg which is 2280 g, so in moles this is $\dfrac{2280}{50.9} = 44.8$ mol. Percentage yield can be calculated as $\dfrac{44.8}{55.0} \times 100 = 81.5\%$. The slight difference in the answers is due to the rounding of the moles of V obtained. Both answers are acceptable.

Question 7

Manganese(ɪv) oxide reacts with warm, concentrated hydrochloric acid to generate chlorine according to the equation:

$MnO_2(s) + 4HCl(l) \rightarrow MnCl_2(s) + Cl_2(g) + 2H_2O(l)$

(a) Calculate the mass of manganese(ɪv) chloride produced when 2.45 g of manganese(ɪv) oxide reacts with an excess of concentrated hydrochloric acid. Give your answer to 3 significant figures. (3 marks)

(b) Calculate the volume of chlorine gas, in cm^3, produced at 100 kPa and 77°C when 2.45 g of manganese(IV) oxide reacts with an excess of concentrated hydrochloric acid. (The molar gas constant R is 8.31 J^{-1}mol^{-1}.) (4 marks)

(a) moles of $MnO_2 = \dfrac{2.45}{86.9} = 0.02819 \, \text{mol}$ ✓

moles of $MnCl_2 = 0.02819 \, \text{mol}$ ✓   (1 : 1 ratio of $MnO_2 : MnCl_2$)

mass of $MnCl_2 = 0.02819 \times 125.9 = 3.55 \, \text{g}$ ✓

ⓔ This is a straightforward reacting mass calculation. The amount, in moles, of MnO_2 is calculated. The ratio of $MnO_2 : MnCl_2$ is used to calculate the amount in moles of $MnCl_2$. Finally, the moles of $MnCl_2$ are multiplied by the M_r of $MnCl_2$.

(b) moles of $Cl_2 = 0.02819 \, \text{mol}$ ✓

$p = 100000 \, \text{Pa}$

$n = 0.02819 \, \text{mol}$

$R = 8.31 \, \text{J K}^{-1}\text{mol}^{-1}$

$T = 350 \, \text{K} \, (77°C = 350 \, \text{K})$

$V = \dfrac{nRT}{p}$ ✓

$V = \dfrac{0.02819 \times 8.31 \times 350}{100000} = 8.199 \times 10^{-4} \, \text{m}^3$ ✓

$V = 820 \, \text{cm}^3$ ✓

ⓔ Common errors in this type of question mainly involve the units used in the ideal gas equation. Pressure (p) must be in pascals (Pa), volume in m^3, n in mol, R in J K^{-1}mol^{-1} and T in K. In this question the pressure is given as 100 kPa, which converts to 100 000 Pa. The volume will be calculated in m^3 and then changed to cm^3 by × 10^6. The temperature is given as 77°C which, when 273 is added, becomes 350 K. Make sure that you can work with these units comfortably.

Bonding

Question 1

(a) State the type of bonding in magnesium chloride. Explain why a lot of energy is needed to melt a sample of solid magnesium chloride. (3 marks)

(b) Explain why a solution of magnesium chloride conducts electricity. (2 marks)

(a) Ionic ✓

There is a strong electrostatic attraction ✓ between the oppositely charged ions ✓.

ⓔ Ionic bonding is between a metal and a non-metal (with some exceptions, for example, beryllium and aluminium chlorides). Since magnesium is a metal ion and chloride a non-metal ion, the bonding is ionic. In your answer you must indicate that there is a strong attraction in the bond, and this requires a lot of energy to break it. The final mark is given for stating what the attraction is between the oppositely charged ions/positive and negative ions/Mg^{2+} and Cl^-. Often this last mark is lost.

(b) Magnesium chloride is an ionic compound and when dissolved in water the ions ✓ are free to move and carry charge ✓.

ⓔ The most common error here is to confuse ions and electrons. Metals and graphite conduct electricity due to the delocalised electrons, which are free to move and carry charge, but molten ionic compounds, ionic compounds dissolved in water and acids all conduct electricity due to the free ions that can move and carry charge. You will lose a mark if you confuse the charged particles that can move.

Question 2

Which one of the following liquids is polar?

A CCl_4 **B** CS_2 **C** C_2H_5OH **D** C_6H_{14}

Answer is C ✓

ⓔ This type of question can be asked in many different ways and the molecules chosen can vary. Again, use your knowledge of shape and polarity to decide which are the non-polar molecules. CS_2 will be non-polar, based on CO_2 being non-polar. All hydrocarbons are non-polar. CCl_4 and CS_2 both contain polar bonds but the bonds are arranged symmetrically so the polarities cancel out, making the molecules non-polar overall. The question could ask, 'which one of the following liquids would be deflected by a charged rod?' The answer and reasoning are the same.

Question 3

Explain why water has a higher melting point than hydrogen sulfide.   (3 marks)

Hydrogen bonds between water molecules ✓ are stronger than van der Waals forces of attraction/permanent dipole attraction in H_2S ✓ so more energy is needed to break the stronger hydrogen bonds ✓.

ⓔ Any question about physical state or melting/boiling points of simple covalent substances is linked to intermolecular forces. It is vital that you can correctly identify a substance as simple covalent and then identify the intermolecular forces that are needed to explain the properties. Water forms strong hydrogen bonds between its molecules as well as van der Waals forces of attraction whereas hydrogen sulfide, H_2S has weaker van der Waals forces of attraction and permanent dipole attractions. The answer must give the type of intermolecular bonding and say which is stronger. Give a full answer, bringing in the idea of the energy required to break the bonds to ensure you gain all the marks.

Question 4

Which one of the following does not have a molecular covalent crystalline structure?

A diamond   **B** ice   **C** iodine   **D** sulfur

Answer is A ✓

ⓔ This question can be approached in two ways as it is a negative question. Highlight the 'not' to remind you. Describe the structure of each molecule and decide which one does not exist as molecular covalent crystals. All are crystalline solids but diamond is giant covalent; the rest are molecular covalent.

Question 5

The halogens are found in group 7 and show a pattern in electronegativity as the group is descended.

(a) (i) Explain what is meant by the term electronegativity.   (1 mark)

(a) (i) Electronegativity is the power of an atom to attract the pair of electrons in a covalent bond ✓.

ⓔ This is a common definition style question and it is important that you include all aspects of the definition to gain the marks. Do not change it, and make sure that you use the proper scientific terms, as often one wrong word can cost you a mark. It is essential that you realise that electronegativity is the atom's attraction for electrons *in* the bond.

(a) (ii) Write an equation to show the formation of 1 molecule of ClF_3 from chlorine and fluorine molecules.   (1 mark)

ⓔ In this question you must realise that the halogens are diatomic molecules, for example, Cl_2 and F_2. Write the correct formulae in the equation $Cl_2 + F_2 \rightarrow ClF_3$ and then balance the equation by putting numbers in front: $Cl_2 + 3F_2 \rightarrow 2ClF_3$.

The question is to show the formation of one molecule of ClF_3, hence you divide the equation by 2 to get 1 molecule of ClF_3.

(a) (ii) $\frac{1}{2}Cl_2 + \frac{3}{2}F_2 \rightarrow ClF_3$ ✓

(b) Name the shape of a dichlorodifluoromethane molecule (CCl_2F_2) and the shape of a chlorine trifluoride molecule (ClF_3). (2 marks)

(b) CCl_2F_2: Tetrahedral ✓

ClF_3: T shaped ✓

ⓔ First calculate the number of electrons around the central atom. C has four outer shell electrons; add on four for the four bonded halogen atoms, gives eight electrons, which is four pairs. There are four bonded pairs due to the four halogens and no bonded pairs. The four bonding pairs of electrons will repel each other equally and result in a tetrahedral shape.

Chlorine has seven outer shell electrons. Adding on the three bonded electrons from fluorine makes ten, which is five electron pairs. Take away the three bonded electrons pairs give two lone pairs. ClF_3 has three bonding pairs of electrons and two lone pairs of electrons. The lone pairs will repel more than the bonding pairs and the shape is T shaped. Trigonal planar is also accepted.

(c) Suggest the strongest type of intermolecular force between CCl_2F_2 molecules. (1 mark)

(c) Dipole–dipole interactions ✓

ⓔ There are three types of intermolecular force. In decreasing order of strength, these are — hydrogen bonds, dipole–dipole interactions and induced dipole–dipole interactions. Between molecules of CCl_2F_2 there will not be hydrogen bonds, as there is no polar hydrogen. There will, however, be polar bonds in the molecules C–Cl and C–F, so there will be dipole–dipole interactions.

(d) BF_3 is a covalent molecule that reacts with an F^- ion to form a BF_4^- ion. Name the type of bond formed when a molecule of BF_3 reacts with an F^- ion. Explain how this bond is formed. (2 marks)

(d) Coordinate/dative covalent ✓

Lone pair of electrons/both electrons on F^- are donated to the BF_3 ✓.

ⓔ BF_3 has three bonding pairs of electrons; F^- has four lone pairs of electrons. The bond between them forms when the fluoride ion donates a lone pair of electrons. Both of the electrons in the bond are supplied by the F^-. This is a coordinate bond.

(e) (i) Label a Be–Cl bond to show the polarity of the bond. (1 mark)

(e) (i) $Be^{\delta+} - Cl^{\delta-}$ ✓

(e) This question is essentially asking about the trends in electronegativity of elements. The more electronegative atom will be δ– and the other will be δ+. Atoms that are further to the right in the periodic table are more electronegative and those further up the periodic table are more electronegative. Cl is much further to the right than Be and is more electronegative, so Cl gets the δ– and Be gets the δ+. Make sure that you write them above the atoms to indicate the polarity.

(e) (ii) Explain why beryllium chloride is a non-polar molecule even though it contains polar bonds. (2 marks)

> **(e) (ii)** Beryllium chloride contains equally polar bonds, which are arranged symmetrically ✓ and so the polarities of the bonds (dipoles) cancel each other out ✓.

(e) This is a common question as many molecules are non-polar even though they contain polar bonds. Molecules like this must contain equally polar bonds, which are arranged symmetrically so that the polarities (dipoles) of these bonds cancel each other out. Common examples are CO_2, CCl_4 and BF_3. Similar molecules like CS_2, CF_4 and BCl_3 would also be non-polar for the same reasons.

Question 6

Which one of A to D in Table 3 gives the correct shapes of the molecules ammonia, water and carbon dioxide?

	Ammonia shape	Water shape	Carbon dioxide shape
A	Tetrahedral	Linear	Bent
B	Pyramidal	Linear	Linear
C	Trigonal planar	Bent	Linear
D	Pyramidal	Bent	Linear

Table 3

> Answer is D ✓

(e) It is important to first write down the number of lone pairs and bonding pairs in each molecule: NH_3 has three bonding pairs and one lone pair; H_2O has two bonding pairs and two lone pairs; CO_2 has two bonding pairs (remember a double bond acts as a single bonding pair). You should learn the table on page 44 and remember the shapes of various molecules for the different lone pair/bonding pair combinations. Water is often wrongly identified as linear and ammonia as trigonal planar. Carbon dioxide is linear due to the equal repulsion of the bonding electrons in the double bonds. Remember that two bonds can be linear or bent; three bonds can be trigonal planar or pyramidal depending on the presence or absence of lone pairs.

Energetics

Question 1

The standard enthalpy change of combustion of ethane, C_2H_6, is $-1560 \, kJ \, mol^{-1}$. A sample of $0.600 \, g$ of ethane was burned completely in air. The heat produced was used to heat $200 \, g$ of water. Calculate the temperature change observed in the water given that the specific heat capacity of water is $4.18 \, J \, K^{-1} \, g^{-1}$. Give your answer to 3 significant figures.

(3 marks)

ⓔ In this question you are calculating ΔT rather than the standard enthalpy change of combustion using ΔT. The same method is used, just reversed.

$1560 \, kJ$ ($1\,560\,000 \, J$) of energy released when 1 mole of C_2H_6 is burned

moles of propane burned $= \dfrac{0.600}{30.0} = 0.0200$ ✓

energy released when 0.02 moles of propane are burned $= 0.0200 \times 1560000 = 31200 \, J$ ✓

$q = mc\Delta T \qquad 31\,200 = 200 \times 4.18 \times \Delta T$

$\Delta T = \dfrac{31\,200}{200 \times 4.18} = \dfrac{31\,200}{836} = 37.3 \, K$ or $37.3 \, °C$ ✓

ⓔ Be careful with calculations like the last one. If you type $31\,200 \div 200 \times 4.18$ into your calculator it will divide $31\,200$ by 200 and then multiply the answer by 4.18. It is good practice to calculate the numerator (top) and denominator (bottom) separately and then do the division or put the whole of the bottom line in brackets in your calculator. All the values given in the question are to 3 significant figures, and the answer is also to 3 significant figures.

Question 2

Sodium nitrate(v) thermally decomposes to form sodium nitrate(III) and oxygen gas, according to the equation:

$2NaNO_3(s) \rightarrow 2NaNO_2(s) + O_2(g)$

The standard enthalpies of formation are given in Table 4.

	$\Delta_f H^{\ominus} \, / kJ \, mol^{-1}$
Sodium nitrate(v), $NaNO_3$ (s)	-466.7
Sodium nitrate(III), $NaNO_2$ (s)	-359.4
Oxygen, O_2 (g)	0

Table 4

(a) Why is the standard enthalpy of formation of oxygen zero? (1 mark)

(b) Calculate the standard enthalpy change of reaction for the thermal decomposition of sodium nitrate(v). (3 marks)

Questions & Answers

ⓔ The majority of enthalpy change questions focus on organic chemistry but enthalpy questions can be applied to an inorganic reaction.

(a) Oxygen is an element in its standard state ✓

(b) $2NaNO_3(s) \rightarrow 2NaNO_2(s) + O_2(g)$

When using enthalpies of formation, the expression below is used:

$\Delta_r H^{\ominus} = \sum \Delta_f H^{\ominus}(\text{products}) - \sum \Delta_f H^{\ominus}(\text{reactants})$ ✓

$\Delta_r H^{\ominus} = 2(-359.4) - 2(-466.7)$ ✓

$= +214.6\,kJ\,mol^{-1}$ ✓

ⓔ This reaction is endothermic overall, as you would expect a thermal decomposition to be. Remember to include the sign (+ or –) in front of the enthalpy change value, as leaving this out can cost you a mark.

Question 3

Propane gas burns in oxygen according to the equation:

$C_3H_8(g) + 5O_2(g) \rightarrow 3CO_2(g) + 4H_2O(g)$

The mean bond enthalpies are given in Table 5.

Bond	Mean bond enthalpy ($kJ\,mol^{-1}$)
C–H	412
C–C	348
O–H	463
O=O	496
C=O	803

Table 5

Which one of the following is the standard enthalpy change of combustion of propene?

A $-198\,kJ\,mol^{-1}$

B $-2050\,kJ\,mol^{-1}$

C $-2398\,kJ\,mol^{-1}$

D $-4034\,kJ\,mol^{-1}$

Answer is B ✓

ⓔ The total enthalpy required to break the bonds, based on the main equation given is: $2 \times 348 + 8 \times 412 + 5 \times 496 = 6472\,kJ$. The energy released when bonds form, based on the main equation is: $6 \times 803 + 8 \times 463 = 8522\,kJ$. The enthalpy change in the equation is $+6472 - 8522 = -2050\,kJ$. The distractors are: A, where 4 moles of O–H bonds are used; C, where only 1 mole of C–C bond is included; and D, where only one O=O bond is broken, as opposed to five.

Kinetics

Question 1

What label is placed on a vertical axis of a distribution of molecular energies?

A energy

B number of molecules

C frequency

D enthalpy

> Answer is B ✓

🄴 The vertical axis is number of molecules. Energy is the horizontal axis and enthalpy would also be a distractor, as enthalpy level diagrams are also used in kinetics.

Question 2

For a gaseous reaction:

$$A(g) + 3B(g) \rightleftharpoons 2C(g) \quad \Delta H = -92 \, kJ \, mol^{-1}$$

Figure 1 shows the distribution of molecular energies in the reaction mixture at 450 °C.

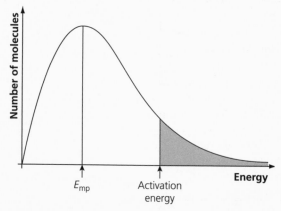

Figure 1

(a) (i) Sketch on the diagram the distribution of molecular energies at 500°C. (1 mark)

(ii) Explain, using Figure 1, why the rate of reaction would be faster at 500°C. (2 marks)

(iii) Explain how the yield of C is affected by increasing the temperature to 500°C. (2 marks)

(b) Suggest why a high pressure is used. (2 marks)

(c) Explain, using Figure 1, how a catalyst increases the rate of reaction. (2 marks)

(d) State the effect, if any, on the most probable energy (E_{mp}) of increasing the concentration of A and B. (1 mark)

(a) (i) The sketch for 500°C should be lower and more to the right ✓

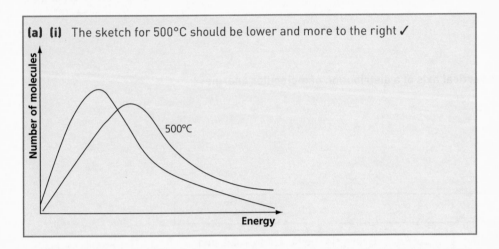

(a) (ii) There will be more molecules above the activation energy ✓ and so more successful collisions ✓.

ⓔ This is a common question and you should understand the link between the shape of the curve and the rate of reaction in terms of the activation energy. Always make sure you try to keep the area under the curve roughly the same. If it is a higher temperature the curve will be more spread out with a lower peak.

(a) (iii) Reaction is exothermic so increasing the temperature moves the position of equilibrium in the direction of the reverse reaction to oppose the change. The position of equilibrium moves from right to left ✓.

The yield of C decreases ✓.

ⓔ Make sure that you explain that the position of equilibrium moves in a direction to oppose the change applied. A question on yield always requires an answer as to whether the yield increases or decreases — make sure that you include an explanation of equilibrium shift in your answer. Just stating that equilibrium moves in a certain direction may not be enough.

(b) There are fewer moles of gas on the right — 4 moles of gas on left and 2 moles of gas on right — so the equilibrium position moves from left to right to reduce volume, that is, to oppose the effect of increased pressure ✓.

There is a higher rate of reaction at increased pressure, as volume is reduced ✓.

ⓔ The effects of temperature, pressure and the presence of a catalyst on rate and on equilibrium are very common. Both the idea of the number of gas moles and the equilibrium position moving from left to right are required for the first mark.

(c) (i)	A catalyst provides an alternative reaction route of lower activation energy ✓.
	More molecules with energy greater than the activation energy, so there are more successful collisions ✓.

ⓔ Again, the reference to the diagram is essential. A catalyst does not affect the position of equilibrium but will increase the rate, so equilibrium is attained more quickly.

(d) E_{mp} increases ✓

ⓔ The only factor that will affect E_{mp} is a change in temperature.

Equilibrium, Le Chatelier's principle and K_c

Question 1

The reaction of hydrogen with iodine is in equilibrium.

$$H_2(g) + I_2(g) \rightleftharpoons 2HI(g) \qquad \Delta H = -10.4\,kJ\,mol^{-1}$$

Which one of the following would move the position of equilibrium from left to right?

A   adding hydrogen iodide

B   decreasing the temperature

C   decreasing the pressure

D   adding a catalyst

Answer is B ✓

ⓔ Adding hydrogen iodide would move the position of equilibrium from right to left. As there are equal number of moles of gas on each side a change in pressure has no effect on the position of equilibrium and a catalyst does not affect the position of equilibrium just increases the rate of reaction.

Question 2

What are the units of K_c for the equilibrium below?

$$N_2(g) + 3H_2(g) \rightleftharpoons 2NH_3(g)$$

A $mol^2\,dm^{-6}$

B $mol\,dm^{-3}$

C $mol^{-1}\,dm^3$

D $mol^{-2}\,dm^6$

Answer is D ✓

ⓔ The expression for K_c for the reaction, is:

$$K_c = \frac{[NH_3]^2}{[N_2][H_2]^3}$$

There is concentration squared on the top of the expression and concentration to the power 4 on the bottom, that is:

$$\frac{conc^2}{conc^4} = \frac{1}{conc^2} = conc^{-2}$$

The units of concentration are $mol\,dm^{-3}$ so $(mol\,dm^{-3})^{-2} = mol^{-2}dm^{-6}$. The other answers are common units of K_c and all are correct. K_c will either have no units or have units that are $(mol\,dm^{-3})^x$ where x is a positive or negative integer. The common mistake here would be to give A as correct, as it has units $(mol\,dm^{-3})^2$, as opposed to the correct answer, which is $(mol\,dm^{-3})^{-2} = mol^{-2}dm^6$.

Question 3

Consider the four gaseous reactions, A to D:

Reaction A	$2P(g) + Q(g) \rightleftharpoons 2R(g)$	$\Delta H = -45\,kJ\,mol^{-1}$
Reaction B	$3S(g) + T(g) \rightleftharpoons 2U(g)$	$\Delta H = +15\,kJ\,mol^{-1}$
Reaction C	$V(g) + W(g) \rightleftharpoons 2X(g)$	$\Delta H = -95\,kJ\,mol^{-1}$
Reaction D	$Y(g) \rightleftharpoons 2Z(g)$	$\Delta H = +127\,kJ\,mol^{-1}$

(a) In which reaction would the position of equilibrium move from right to left with a decrease in pressure? (1 mark)

(b) For which equilibrium reaction would K_c have no units? (1 mark)

(c) State the effect, if any, on the position of equilibrium in reaction B of increasing the temperature. (1 mark)

(d) State the effect, if any, on the value of K_c for reaction A of increasing the pressure at constant temperature. (1 mark)

(a) Reaction D ✓

ⓔ A decrease in pressure moves the position of equilibrium to the side with fewer gas moles to oppose the change. Reaction D is the only reaction that has fewer gas moles on the left (one on the left and two on the right). Reaction A has three on the left and two on the right; reaction B has four on the left and two on the right; reaction C has two on the left and two on the right. The position of equilibrium for reaction C would be unaffected by a change in pressure as there are equal gas moles on each side of the reaction.

(b) Reaction C ✓

ⓔ When there are equal number of moles of reactants and products the concentrations of the products over the reactants cancel out to give no units. The only reaction with equal moles of reactants and products is C. The units of K_c for the other reactions would be: A = $mol^{-1}dm^3$; B = $mol^{-2}dm^6$; D = $mol\,dm^{-3}$. Always check the others to make sure you are correct.

(c) (left) to right ✓

e Reaction B is endothermic in the forward direction. We know this because ΔH is positive. An increase in temperature moves the position of equilibrium in the direction of the endothermic reaction to absorb the heat. You would write this type of answer if you were asked to explain why the position of equilibrium moves from left to right when temperature is increased.

(d) no effect ✓

e The only factor that has any effect on the value of the equilibrium constant K_c is a change in temperature. As the temperature remains constant, the value of K_c will remain the same.

Question 4

For the reaction:

$$2A(g) + B(g) \rightleftharpoons C(g) + D(g)$$

(a) Write an expression for the equilibrium constant, K_c for this reaction. (1 mark)

(b) At 400 K, K_c for this reaction is $1.45 \times 10^{-2}\,mol^{-1}\,dm^3$. In one equilibrium mixture the concentration of A is $0.452\,mol\,dm^{-3}$; B is $0.102\,mol\,dm^{-3}$ and C is $0.277\,mol\,dm^{-3}$. Calculate the concentration of D in the equilibrium mixture. (3 marks)

(c) (i) 0.2 moles of A and 0.1 moles of B were mixed in a container of volume $5\,dm^3$ at 700 K. The mixture was allowed to reach equilibrium and was found to contain 0.15 moles of A. Calculate a value for the equilibrium constant, K_c, at this temperature. (4 marks)

(ii) Using the value for K_c you obtained in (c) (i) and the value given in (b), state whether the reaction is exothermic or endothermic and explain your answer. (2 marks)

(a) $K_c = \dfrac{[C][D]}{[A]^2[B]}$ ✓

e Always take care when writing expressions for the equilibrium constant, K_c, as it is products over reactants. Square brackets are essential as these represent concentration. Remember to raise the concentrations to the powers of the balancing numbers given in the equation for the reaction.

(b) $[D] = \dfrac{K_c \times [A]^2 [B]}{[C]}$ ✓

$[D] = \dfrac{1.45 \times 10^{-2} \times (0.452)^2 (0.102)}{(0.277)}$ ✓

$[D] = 1.09 \times 10^{-3}\,mol\,dm^{-3}$ ✓

e Rearranging a mathematical expression is an important skill. Being able to rearrange a K_c expression to calculate the concentration of a reactant or product present at equilibrium is essential in answering this question and the rearranged expression is often worth a mark. If [A] was the required concentration, then

$[A]^2 = \dfrac{[C][D]}{K_c \times [B]}$. Be careful when calculating expressions like this as the bottom line should be in a bracket or calculated separately, then divided.

(c) (i)		2A	+	B	⇌	C	+	D	
Initial moles		0.2		0.1		0		0	
Reacting moles		−0.05		−0.025		+0.025		+0.025	
Equilibrium moles		0.15		0.075		0.025		0.025	✓
Equilibrium concentration		0.03		0.015		0.005		0.005	✓

Table 6

$$K_c = \frac{[C][D]}{[A]^2[B]} = \frac{(0.005)(0.005)}{(0.03)^2(0.015)} \checkmark$$

$$= 1.85 \checkmark \ (\text{mol}^{-1}\,\text{dm}^3)$$

e This is a standard question and it is important that you can manipulate the figures given to determine the moles present at equilibrium. The key point of information that was given was 0.1500 mol of A present at equilibrium. As there were 0.2 mol of A present initially, 0.05 must have reacted. As 2 moles of A react with 1 mol of B in the equation, 0.025 mol of B must have reacted leaving (0.1−0.025) = 0.075 mol of B. 0.025 mol of C and D were both formed. The equilibrium moles are then divided by 5 (as volume is 5 dm³) to calculate the equilibrium concentration. These values are inserted into the K_c expression to calculate K_c. Try it on your calculator. If you type $0.005 \times 0.005 \div 0.03^2 \times 0.015$ you will get 4.17×10^{-4}. However, if you type $0.005 \times 0.005 \div (0.03^2 \times 0.015)$, you will get 1.85.

(ii) endothermic ✓

K_c increases with an increase in temperature ✓

e K_c is only affected by changes in temperature, and an increase in temperature will increase the value for K_c if the forward reaction is endothermic as there will be less reactants and more products. Remember that although an increase in pressure in this reaction would move the position of equilibrium from left to right (as there are 3 moles of gas on the left and 2 on the right), the concentrations after the increase in pressure will still give the same value of the equilibrium constant, as long as temperature remains the same. This can be a sticking point for some students.

Oxidation, reduction and redox reactions

Question 1

Explain, in terms of oxidation states, why the reaction shown is described as a redox reaction. (3 marks)

$$6FeSO_4 + 3Cl_2 \rightarrow 2Fe_2(SO_4)_3 + 2FeCl_3$$

Iron/Fe is oxidised from +2 (in $FeSO_4$) to +3 (in $Fe_2(SO_4)_3$) ✓.

Chlorine is reduced from 0 (in Cl_2) to –1 (in $FeCl_3$) ✓.

Redox is oxidation and reduction occurring in the same reaction ✓.

ⓔ Most answers will achieve the final mark but many will confuse the calculation of oxidation states. Remember that the sulfate ion is SO_4^{2-} so the iron (Fe) in $FeSO_4$ has an oxidation state of +2 and the iron (Fe) in both $Fe_2(SO_4)_3$ and $FeCl_3$ has an oxidation state of +3. All compounds have an overall oxidation state of 0. All elements, even diatomic ones like chlorine, Cl_2, have an oxidation state of 0. The balancing numbers in the equation do not affect the oxidation state. The sulfate ion, SO_4^{2-} does not change so the oxidation state of the sulfur or oxygen also does not change.

Question 2

For the following half-equation, x, y and z are the numbers of moles of water, hydrogen ions and electrons required, respectively.

$$SO_2 + xH_2O \rightarrow SO_4^{2-} + yH^+ + ze^-$$

Which one of A–D in Table 7 is correct?

	x	y	z
A	1	2	2
B	2	2	2
C	2	2	4
D	2	4	2

Table 7

ⓔ To answer this question, look at the oxygen content from left to right. Two oxygens are gained so $2H_2O$ are required on the left to balance the oxygen. This gives four hydrogens on the left and so $4H^+$ are needed on the right. Even at this stage the answer is D, but it is a good idea to check the charges and balance the electrons in case you have made a mistake with the oxygen and hydrogen. The charge on the left is zero whereas the total charge on the right is +2 (–2 for the SO_4^{2-} and +4 for the $4H^+$). This mean $2e^-$ are required on the right to balance the charge. Practise balancing these equations but watch out for the dichromate(VI) one, where the ratio of dichromate to Cr^{3+} is $1:2$.

Answer is D ✓

Question 3

Using the half-equations shown below:

$MnO_4^- + 8H^+ + 5e^- \rightarrow Mn^{2+} + 4H_2O$

$2Br^- \rightarrow Br_2 + 2e^-$

write an ionic equation for the oxidation of bromide ions using acidified potassium manganate(VII).

(2 marks)

ⓔ Before tackling this question it is important to remember that an ionic equation does not contain electrons. You must combine an oxidation and reduction half-equations, multiplying each if necessary so that the electrons cancel out on each side. Remember for more complex ionic equations you may have to cancel out H^+ ions and H_2O as well to simplify the final equation. In this example multiply the manganate(VII) half-equation by 2 and multiply the bromide half-equation by 5. Then add the equations together by writing down everything on the left of both arrows then putting an arrow before writing down everything on the right of the arrows. Then cancel out anything that appears on both sides (in this case $10e^-$) and you have the finished ionic equation.

$2MnO_4^- + 16H^+ + 10e^- \rightarrow 2Mn^{2+} + 8H_2O$

$10Br^- \rightarrow 5Br_2 + 10e^-$

$2MnO_4^- + 16H^+ + 10e^- + 10Br^- \rightarrow 2Mn^{2+} + 8H_2O + 5Br_2 + 10e^-$

Ionic equation:

$2MnO_4^- + 16H^+ + 10Br^- \rightarrow 2Mn^{2+} + 8H_2O + 5Br_2$ ✓✓

Knowledge check answers

1 Fe

2 Zn

3 **a** $Si(g) \rightarrow Si^+(g) + e^-$ **b** $Mg^+(g) \rightarrow Mg^{2+}(g) + e^-$

4 Decreases. This is because there is an increase in atomic radius, there is more shielding of the outer electron from the nuclear charge due to increased shells and as a result there is less nuclear attraction for the outer electron.

5 Argon has a larger nuclear charge than sodium, and the shielding is similar hence it has the smaller atomic radius and the outer electron is held closer to the nucleus by the greater nuclear charge. Sodium's outer electron is not held as tightly due to a smaller nuclear charge.

6 $(NH_4)_2Cr_2O_7$

7 0.0151 mol

8 1.51×10^{22} mol

9 5.34 dm³

10 $C_3H_2NO_2$

11 $HgCl_2$

12 7

13 $Mg_3N_2 + 6H_2O \rightarrow 3Mg(OH)_2 + 2NH_3$

14 0.902 g

15 74.1%

16 nitric acid/sulfuric acid/hydrochloric acid

17 pink to colourless

18 0.0987 mol dm⁻³

19 $x = 2$

20 32.7%

21 One electron from carbon shares with one electron from hydrogen forming a shared pair of electrons.

22 $AsCl_3$ is pyramidal and Cl_3^+ is bent.

23 The electron pair is not shared equally but is displaced more towards the more electronegative oxygen.

24 **a** dipole–dipole interactions **b** hydrogen bonds
   c induced dipole–dipole interactions/van der Waals forces

25 **a** hydrogen bonds **b** dipole–dipole interactions

26 The molecules increase in size (greater M_r) with more electrons, and so increased van der Waals forces between molecules.

27 Phosphine cannot form hydrogen bonds with water, whereas ammonia, due to the polar nitrogen, can form hydrogen bonds with water.

28 In graphite much energy is needed to break the many strong covalent bonds between the carbon atoms. To melt phosphorus less energy is needed to break the weaker van der Waals forces between the molecules.

29 Change in enthalpy

30 298 K 100 kPa

31 m = mass (g)
   c = specific heat capacity ($J K^{-1} g^{-1}$)
   ΔT = change in temperature (K or °C)

32 $-128 \, kJ \, mol^{-1}$

33 $-1367 \, kJ \, mol^{-1}$

34 H–I

35 A catalyst provides an alternative reaction route of lower activation energy.

36 vertical = number of molecules
   horizontal = energy

37 Equal numbers of moles of gas on both sides of the equation.

38 Iron

39 Concentration of reactants and products remain constant.
   The rates of the forward and reverse reactions are equal.

40 $K_c = \dfrac{[NO_2]^2}{[N_2O_4]} \, mol \, dm^{-3}$

41 $1.85 \, mol^{-2} \, dm^6$

42 2.25

43 +3

44 $SO_4^{2-} + 8H^+ + 6e^- \rightarrow S + 4H_2O$

45 O changes from –1 in H_2O_2 to –2 in H_2O
   I changes from –1 in KI to 0 in I_2
   Redox is oxidation and reduction occurring simultaneously in the same reaction
   $H_2O_2 + 2H^+ + 2e^- \rightarrow 2H_2O$

Index

Note: page numbers in **bold** refer to key term definitions.

Index